More Words of Science

ISAAC ASIMOV

Decorations by William Barss

Houghton Mifflin Company Boston 1972

To Robyn Joan
best and sweetest of daughters

Books on words by
ISAAC ASIMOV

Words of Science
Words from the Myths
Words in Genesis
Words on the Map
Words from the Exodus
Words from History
More Words of Science

501.4
As4m
82697
Mar. 1973

LIBRARY OF CONGRESS CATALOG CARD NUMBER 79-187422
ISBN 0-395-13722-5
PRINTED IN THE UNITED STATES OF AMERICA
FIRST PRINTING W

Introduction

My book Words of Science was published in 1959. It was a book in which I told something about the derivation and use of about 1500 words used in science: from words as simple as *oil*, *cell*, and *line* to some as seemingly complicated as *archeozoic* and *elasmobranchii*.

The list was certainly not exhaustive and there remained many words that I had not mentioned. It was always in my mind to write a second collection of such words someday, a collection that would include words such as *corona*, *entropy*, and *thermodynamics*, which I had omitted from the first book.

Then, as the years passed, something else impressed itself upon me forcefully. Science expands its scope steadily from year to year and has been doing so, as it happens, at a continuously increasing rate. And so does its vocabulary.

The words of science multiply in two ways. First, some old words of the English language gain a new connotation. Consider the word *pollution*, for instance. It has been with us for centuries, meaning the introduction of defilement or impurity of any sort. One's mind and soul could be polluted by evil, for example.

Nowadays, however, *pollution* has come to mean one specific thing — the introduction of wastes into the environment at a rate faster than they can be removed naturally. The word now represents something of overwhelming importance to ecologists. For that matter, *ecology* itself is an old word that lived unnoticed in the dictionary until it came into prominence almost overnight. Now it is a household expression.

Consider also such ordinary words and phrases as *fallout, black hole,* and *big bang,* which now have definite meanings in the scientific world.

Second, there are new words in science because there are things and concepts in science that are literally new and unprecedented and that require the actual manufacture of new words to symbolize them.

I could not include the word *quasar* in my 1959 book for the simple reason that in 1959 the mysterious, distant objects in the sky which are now represented by that word had not yet been discovered! For similar reasons, I could not include *laser,* or *quark,* or *pulsar,* or *lawrencium,* or *transfer RNA.*

So here we are with MORE WORDS OF SCIENCE, in which some of the words are new only because I treat them in the second book after having omitted them in the first; others, because while a particular word is old, the scientific usage is new or has only become prominent in recent years; and still others, because they were only invented in the last dozen years.

In this book, by the way, I have decided to avoid the cross-references I used in WORDS OF SCIENCE. Cross-references have the advantage of allowing an author to move faster and to skip details in a particular essay by making the reader turn to another page for more information. They have the disadvantage of occasionally annoying a reader who may not be in the mood to turn to another page.

This time, then, I tried to make each essay as complete in itself as I could in the one page allotted me. For that reason, you will find occasional partial overlapping. But then you can't have everything, can you?

Ablation

IN SOME WAYS, a spacecraft is in greatest danger when it is within a hundred miles of the surface of Earth. It is then not in space, but in Earth's atmosphere, which heats it by friction.

Moving away from Earth is not so bad. The spacecraft moves upward very slowly at first, so that there is little friction with the air. As the rocket engines drive it faster and faster, the air it penetrates is thinner and thinner and offers less and less friction. By the time the spacecraft is moving really fast, the air about it is so thin that there is no danger.

Returning to Earth and reentering the atmosphere is different. The space capsule, with its human passengers, is moving faster and faster as it approaches Earth, thanks to the steady pull of Earth's gravity. Eventually, the air will be dense enough to allow the use of parachutes to slow the fall, but by that time, frictional heat will have been deadly.

To avoid this, the capsule is shaped like a cone. The blunt bottom end hits the atmosphere, pushing the air ahead of it and forming a shock wave that cushions the capsule and acts like a spring, slowing it down. This takes care of from ninety to ninety-five per cent of the capsule's motion.

What's more, the blunt end of the capsule is covered with a layer of resin, strengthened with glass fibers. The resin chars, melts, and flakes away, carrying heat with it. The charred resin left behind is black and radiates additional heat.

This system of removing heat by flaking away a layer is an effect called *ablative cooling*. The resin undergoes *ablation*, which is from Latin words meaning "to carry away." The word was once used to mean the removal of part of the body by surgery, but has now joined the world of space flight.

1

Acetylcholine

In 1861, the German chemist Adolph Strecker isolated a hitherto unknown nitrogen-containing compound from liver bile. He named it *choline* from a Greek word meaning "bile."

In 1914, the English biologist Henry H. Dale isolated a substance with a molecule composed of two parts linked in a fashion called an *ester linkage*. (The word *ester* was manufactured from the first and last syllables of the German name for the best-known compound of this type, back about 1880). Dale separated the two parts and identified one as choline and the other as a well-known compound, acetic acid. He therefore named the combination *acetylcholine* and observed that it had a strong effect on various organs.

In 1921, a German physiologist, Otto Loewi, discovered that nerve endings liberated small traces of a chemical that activated nearby cells and allowed nerve impulses to leap the intervals between one nerve cell and the next. He called the chemical substance *Vagusstoff*, which is German for "vagus material," because he obtained evidence for its existence by stimulating the vagus nerve.

Since *Vagusstoff*, by stimulating nerves, produced effects on organs similar to that produced by acetylcholine, Dale suggested that *Vagusstoff was* acetylcholine and by 1933 had proved his case.

Acetylcholine remains in existence at nerve endings only long enough for the impulse to cross from one cell to another and then is broken up to acetic acid and choline. This breakup is hastened by the presence of an enzyme.

Enzymes are usually named for the type of reaction they hasten, with an "-ase" suffix added. Thus, the enzyme which brings about the breakup of the ester linkage in acetylcholine is named *cholinesterase.*

Actinide

SINCE THE 1870s, it has been known that the chemical elements exist in families. Tables of elements (periodic tables) have been prepared which present those families in rows down, or lines across. Thus, the 57th element, lanthanum, is followed by fourteen elements (58 to 71, inclusive) that are similar to lanthanum and make up the rare earth family. This is usually presented as a horizontal line in the table. Lanthanum is also part of a family of elements arranged in a vertical row. Above it are scandium (21) and yttrium (39).

In 1899, element 89 was discovered and named *actinium* because it gave off radioactive rays and the Greek word for "ray" is *aktinos*. Actinium fits under lanthanum in the periodic table and has properties like scandium, yttrium, and lanthanum.

Was actinium followed by a series of similar elements as lanthanum was? At first chemists thought not. They felt the rare earths to be a special case and considered the element after actinium (which was thorium, element 90) to be like hafnium (element 72), which appeared in the table after all the rare earths were over and done with.

Until 1940, however, only three elements beyond actinium were known, so the evidence was inconclusive. In that year chemists began to construct atoms of higher elements in the laboratory — neptunium (93), plutonium (94), and so on. With more elements to study, new evidence appeared, and in 1944, the American chemist Glenn T. Seaborg was able to show that the elements after actinium made up a second series very like the rare earths.

A name was needed to distinguish one series from the other and it was decided to name each series after its first member. The first series of rare earths were the *lanthanides* after lanthanum, and the newly discovered series were, naturally, the *actinides*.

Adamantane

THE ANCIENT GREEKS imagined a material so hard that it could not be scratched or dented in any way. They called it *adamas*, which in their language meant "untamable."

There is nothing that is infinitely hard, but in the form of *adamant*, the expression came to be applied to any metal that was particularly hard. Finally, it came to be applied to a glassy form of carbon, rarely found, which was harder than any other substance found in nature. If you drop the initial *a* and distort the rest of the word slightly, you end up with *diamond*, which is what this hardest substance is now called in English.

The carbon atoms that make up diamond are very small and can approach each other quite closely. What's more, they do it in so symmetrical a fashion that each carbon atom is closely approached by four others at equal distances from each other. It is the equal, tight grip that each carbon atom has on its neighbors that makes diamond so hard.

Organic compounds are made up of chains and rings of carbon atoms, but generally, the carbon atoms are not arranged as in diamond. Instead, they are arranged in lines or rings that are less symmetrical than the carbon distribution in diamond.

Tiny quantities of a compound with a molecule containing ten carbon atoms were first prepared in large amounts in 1957. The ten carbon atoms were found to be arranged in three interconnected rings, with a distribution exactly as one would find in ten neighboring carbon atoms in diamond. This molecule (which carried sixteen hydrogen atoms attached to its carbons) was particularly stable and was named *adamantane*. The "-ane" suffix is used by chemists for compounds made up of carbon atoms attached to all the hydrogen atoms they can hold. It turns out that modified adamantane compounds may be useful in medicine, since they block virus action in some cases.

4

Aflatoxin

A CERTAIN FUNGUS attacks rye, causing infected grains to darken and curve into the appearance of a rooster's spur, which, in medieval French, was called an "argot." In English, this has been corrupted to *ergot*, which is the common name given to the fungus. This fungus produces certain compounds that have powerful effects on the body even in small quantities. It causes contractions of the uterus, for instance, so that medieval midwives used it sometimes to ease difficult childbirths. It was a dangerous remedy, however, for too much of it produces a serious and even fatal condition known as *ergotism*. Every once in a while, there are reports of an epidemic of such a disease among humans or animals that have been eating contaminated rye.

Nor is ergot the only fungus capable of producing compounds which, in small quantities, have profound and usually unpleasant effects on men. These compounds are called *mycotoxins*. The prefix "myco-" is from a Greek word meaning "mushroom," which is the most familiar fungus because it is so large. (Most fungi are microscopic organisms.) *Toxin* is from the Greek word for "bow," since arrows were so frequently poisoned in warfare. A mycotoxin is therefore a fungus poison.

In 1960, an epidemic of serious liver disease was traced to the eating of moldy peanuts. The mold concerned was a very common one of the *Aspergillus* group. This name was used because such molds were made up of strands swollen at one end so that they looked like tiny versions of a device (aspergillum) used in churches. The aspergillum had small holes in its swollen end and was used to sprinkle holy water, its name coming from the Latin word for "sprinkle." The particular variety of mold was *Aspergillus flavus* (the latter word from a Latin word meaning "yellow," because of its color). From the initials *A. fla.* the poisons it produced were called *aflatoxins*.

5

Airglow

ON A CLEAR MOONLESS NIGHT, there can sometimes be detected a very faint general luminosity in the sky, which seems to originate in the upper atmosphere. It is called *airglow*. It is there both day and night, but it is so much more intense at night, and so much more easily detected then, that it is sometimes called *nightglow*.

The nature of the airglow is revealed by a careful analysis of its light as photographed from the ground and by rockets sent into the upper atmosphere.

The upper atmosphere, it turns out, is rich in atomic oxygen — that is, single oxygen atoms rather than the oxygen molecules of the lower atmosphere, which are made up of two joined oxygen atoms each.

Apparently, the ultraviolet light of the sun is absorbed by oxygen molecules in the upper atmosphere, and as a result the oxygen atoms in the molecule gain too much energy to remain together. They split apart. Every once in a while, two oxygen atoms collide, give off their excess energy as a tiny flash of visible light, and combine again. During the day, the reunion is less frequent than the split-up, and by the time night falls, there is a large supply of atomic oxygen in the upper atmosphere. Through the night this combines, producing the airglow.

Other upper-atmosphere phenomena are the very thin clouds one can see at heights of eighty kilometers (fifty miles) or more. They are most easily seen just after sunset in high latitudes in the summer, when they seem to shine very faintly against the darkening sky. They are *noctilucent clouds* (from Latin words meaning "night shining"). Rockets have brought back particles of matter from those regions and chemical analysis has shown the presence of nickel. It may be, then, that noctilucent clouds are made up of very fine dust particles produced by tiny meteors that flash into the upper atmosphere, heat, and disintegrate into powder.

Aleph-One

WHEN WE COUNT, we match one set of objects with another. We point to a set of objects and say "one" when we point to the first, "two" when we point to the second, and so on, matching the set we are counting to the set of whole numbers.

If two sets can be made to match exactly, we know the number representing one set is equal to the number representing the other. For instance, the set of even numbers: 2, 4, 6, 8 . . . can be matched evenly with the set of whole numbers: 1, 2, 3, 4 . . . Each even number is matched with the whole number equal to half itself. There would be a different whole number for each even number. We can therefore say that the set of even numbers is equal to the set of whole numbers. This seems paradoxical since the set of whole numbers includes odd numbers as well as even numbers and we should think there would be twice as many whole numbers as even numbers.

In 1895, the German mathematician Georg Cantor worked out the mathematics of *infinite sets* (whole numbers and even numbers continue endlessly and are therefore infinite sets) and showed that they do not follow the ordinary rules of arithmetic. He called the numbers representing endless sets *transfinite numbers* (from Latin words meaning "beyond the end").

Transfinite numbers are symbolized by *aleph*, the first letter of the Hebrew alphabet. The number of whole numbers is the lowest transfinite number, *aleph-null*. Any infinite set which can't be matched with the set of whole numbers has a higher transfinite number. The set of points in a line is higher and may be equal to *aleph-one*, the second lowest transfinite number. In the late 1960s, however, the American mathematician Paul J. Cohen showed that it wasn't possible either to prove or disprove the statement that the set of points in a line is equal to aleph-one.

7

Analog Computer

MOST SIMPLE MATHEMATICAL PROBLEMS deal with actual numbers: $2 + 3$; 18×6; and so on. The answers are exact. If you manipulate separate objects, like pebbles or wheel notches or electricity pulses, you can get answers with what is called a digital computer.

Some mathematical problems, however, don't deal with exact numbers. The diagonal of a square is equal to 1.4142 . . . (an endless decimal) times the sides. Or you might have to determine the direction in which to point a gun, firing a shell traveling at such and such a velocity, in order to hit an object moving in a given direction at a given velocity, which may be changing, making allowance for the wind, for the curvature of Earth, and so on — where none of the values are absolutely exact.

In that case, it might be handy to make use of something that varies just the way the conditions in your problem do, but in a way that is easier to handle. A slide rule, for instance, gives approximate answers to numerical problems by having each number represented by a length. These lengths can be manipulated by sliding a piece of wood back and forth, which, in this way, imitates the arithmetical behavior of the numbers and produces the answer. Because lengths are analogous to numbers here, the slide rule is a simple *analog computer*.

In electronic devices, it is current strength that can be treated as analogous to numbers. This is made to vary very quickly in accordance with conditions designed to resemble the variations of the numbers in the problem. The final current strength gives the answer desired, if the data fed into the machine is accurate in the first place.

Both analog computers and digital computers do what the human brain can do, but many millions of times faster. So far, though, they only work as manipulated by human beings and are therefore still only supercomplex slide rules.

Anechoic Chamber

SOUND CONSISTS of waves of compression in the air, moving outward from their starting point at about 740 miles per hour. If a sound wave hits a hard obstacle, it is reflected. If the obstacle is far enough away, it takes enough time for the sound to go there and bounce back to produce a second sound you can hear separately. This is an *echo* from a Greek word meaning "sound."

At shorter reflection distances, the echo is not heard separately, but makes the original sound last longer, and may cause it to get louder and softer in rapid succession as echo after echo returns. The result is *reverberation*, from a Latin word meaning "to strike back." Too much reverberation can be disturbing, and in a poorly designed hall it can make speech unintelligible and music discordant.

Sound can be absorbed. If the waves strike a soft material with small openings, such as clothing, drapery, or carpeting, the sound waves lose themselves in the openings and are not reflected. The study of room design to cut down undesirable reverberation is called *acoustics* from a Greek word meaning "to hear." Sometimes this word is applied to the study of sound generally.

We are used to a certain amount of reverberation, for sound always echoes from walls, floors, rocks, trees, and ground. That is why sound seems strange in rooms that have been designed with surfaces so broken up and so lined with absorbing material as to allow virtually no sound reflections at all. Such a room is an *anechoic chamber* from Greek words meaning "no echoes." Sounds made in the room die out so quickly in the absence of echoes that such chambers are also called *dead rooms*. Anechoic chambers are used to adjust microphones and other acoustical devices, when one wishes to avoid the confusing existence of even the slightest echo or reverberation.

Aneurysm

THE HUMAN HEART beats constantly at a rate of 60 to 80 times a minute throughout a long life that may last over a century. In a hundred years of faithful labor, it will beat some four billion times and pump approximately 600,000 tons of blood.

With each beat, the heart ejects about 130 cubic centimeters of blood, and most of this goes into the aorta at a speed of 40 centimeters per second. The surge of blood subjects the aorta, the largest and most important artery in the body, to a periodic strain. Fortunately, the aorta and other arteries have thick, elastic walls which expand as the blood tumbles in and contract again as it passes by. To a certain extent, the arteries mimic the heartbeat, keeping time and allowing for whatever delay is required for the blood surge to reach them. This arterial beat is called the *pulse*, from a Latin word meaning "to beat," and it can be felt easily where arteries are near the skin, notably in the wrist.

If, for any reason, the wall of an artery is damaged or thinned so that it expands too much when the blood surge enters and contracts too little afterward, the result is an *aneurysm*, from a Greek word meaning "widening." Such a flaccid, too wide artery remains in momentary danger of rupture with possibly fatal results. Since the aorta gets the first brunt of the blood surge, it is most subject to aneurysms and the condition is most dangerous there.

Little could be done about aneurysms until recent times. In 1948, the American surgeon Michael E. De Bakey, working at Baylor University in Houston, began to operate on aneurysms. He cut out the affected portions of the vessels and replaced them with less important and unaffected vessels from other parts of the body. Eventually, he replaced them with dacron tubes. Thousands of certain deaths, including that of the Duke of Windsor, have been prevented in this way.

Antideuteron

In the 1930s, physicists became convinced that every subatomic particle had a twin (*antiparticle*) that was opposite to itself in some key property. The first such antiparticle to be discovered was the positron, which was just like an electron but had a positive electric charge the precise size of the electron's negative electric charge. The positron is sometimes called the *antielectron*.

Again, there is the proton, which has a positive electric charge, and the *antiproton*, which has all the other properties of a proton, but has a negative electric charge. The neutron has no electric charge at all, but has a magnetic field pointing in a particular direction. The *antineutron* has a magnetic field pointing in the other way.

Our part of the universe is made up entirely of particles. Antiparticles made in the laboratory soon react with their opposite numbers and disappear. If there were a section of the universe in which only antiparticles existed, it seems certain they would come together in the same way particles do to form whole atoms and molecules made up of antiparticles alone. They would make up *antimatter*.

But is this just theory, or can we find evidence for it? The simplest way to put antiparticles together is to join an antiproton and an antineutron. A proton and a neutron in conjunction make up the nucleus of a type of hydrogen atom twice as massive as ordinary hydrogen (whose atomic nucleus contains only a proton and nothing more). The proton-neutron hydrogen is called *deuterium*, from a Greek word meaning "second," because its nucleus contains a second particle. The nucleus alone, the proton-neutron combination, is a *deuteron*. An antiproton-antineutron combination would therefore be an *antideuteron*. In 1965, American physicists actually produced such antideuterons, and for the first time antimatter (the simplest possible variety) was formed.

11

Antiparticle

THE WORD *particle* is from a Latin word meaning "little part," so that a particle is a very small piece of matter.

Until the 1890s, chemists thought that atoms were the smallest possible pieces of matter, incapable of being broken into anything smaller. The very word *atom* is from a Greek term meaning "unbreakable."

Then in the 1890s, it was found that the atom was made up of still smaller objects. The first of these to be discovered, carrying a negative electric charge, was called an *electron*. Another, considerably larger than the electron and carrying a positive electric charge, was named *proton*. In 1930, an object the size of a proton but with no electric charge at all was discovered and named *neutron*. Since then, dozens of objects smaller than atoms have been discovered. To give them all a name in common, they are called *subatomic particles*.

In 1928, an English physicist, Paul A. M. Dirac, presented theoretical arguments for supposing that for every subatomic particle there should be another that was opposite in properties. For instance, there should be a particle opposite to the electron, with a positive electric charge but otherwise just like it. Such an opposite-of-the-electron particle was discovered in 1932 by the American physicist Carl D. Anderson and was named *positron* because of its positive charge.

It was not till 1955, though, that the opposite of the proton (just like it, but with a negative electric charge) was discovered. Instead of being given a special name it was called an *antiproton*, for the Greek prefix "anti-" means "opposite to."

Physicists went on to discover opposites to all the subatomic particles that could have been in theory. The general name they gave these opposites was *antiparticles*.

Apollo, Project

THE SUCCESS of the American space effort in terms of the one-man orbital flights of Project Mercury and the two-man orbital flights of Project Gemini meant the next effort would surely be the quarter-million-mile leap across space to the moon. In 1960, President John F. Kennedy stated that it was an American aim to get a man to the moon and bring him back to Earth safely by the end of the decade. By the time Project Gemini was concluded, most of the decade was gone.

Since Projects Mercury and Gemini both had names drawn from mythology, it seemed reasonable to use mythology for the moonlanding project as well. The moon deities were all females, however, and a masculine name was desired.

The mythological beings associated with the sun were masculine and the best known was Apollo. The moonlanding program therefore became *Project Apollo*. It was fitting, for Apollo drove the golden, flaming chariot of the sun, while the three men who were to maneuver the space vessel to the moon were driving a fiery rocket exhaust. It was almost as though the myth were coming to life.

There was something ill-omened about driving the chariot of the sun. Phaethon, the son of the sun god, once tried to guide the chariot. In his inexperienced hands, the chariot went off course and endangered the whole world so that Zeus was forced to kill Phaethon.

In January 1967, three astronauts, testing an Apollo capsule, died tragically and Phaethon-like in an accidental fire. Nevertheless, on July 20, 1969, the *Apollo 11* was guided safely to the moon, and Neil A. Armstrong became the first man to set foot on another world. He and his comrades came safely back to Earth and President Kennedy's goal was met, though he himself had died at the hands of an assassin years before and was not present to witness the great day.

13

Appestat

THERE ARE TWO chief factors to weight increase under normal conditions — the amount of calories that comes in as food and the amount that goes out by way of physical activity. A plump person may eat no more than a skinny one, but may be considerably less active.

Many people keep their weight steady year after year just by eating when they are hungry. If they have been very active, they get hungry sooner and stay hungry longer. In that way they take in enough calories to balance the work or exercise they have been doing. If they have been taking it easy for quite a while, they are less hungry and are more quickly filled.

There is some device in the body which controls the appetite in such a way as to keep food intake matched to energy output. This is similar to the way in which a *thermostat* (from Greek words meaning "heat-stationary") controls a furnace so as to keep the temperature steady. The appetite-controlling device is called, by analogy, the *appestat.*

It seems likely that in plump people, who gain weight easily, the appestat is set higher than it should be. They get hungry too soon and stay hungry too long, just as a house might have its thermostat set too high so that it overheats. Plump people find it agony to try to diet, since all the while they are cutting down on food, the appestat is constantly signaling for more.

The appestat seems to be located in a portion of the brain called the *hypothalamus* ("beneath the thalamus"). *Thalamus* is from a Latin word for a kind of room, since the ancient Romans thought this brain portion was hollow like a room. When the hypothalamus of a laboratory animal is damaged, the appestat seems to be shoved high. The animal begins to eat voraciously and soon gets enormously fat.

Astrochemistry

MANY GALAXIES, including our own, contain vast, thin clouds of gas and dust, which astronomers assume to be composed of the common varieties of atoms making up the universe — hydrogen, helium, neon, oxygen, carbon, and nitrogen.

The atoms are spread out so thinly in these galactic clouds that in their random movements, they would be expected to approach one another only rarely. For that reason, astronomers expected to find little, if any, atom combinations. Yet in the 1930s, astronomers could tell from the light absorption of such clouds that there might be small quantities of carbon-hydrogen (CH) and carbon-nitrogen (CN) combinations.

Once radio waves from the heavens began to be studied and analyzed, in the 1940s and afterward, however, a sharper tool was available. In 1963, for instance, radio waves from certain interstellar clouds were analyzed and found to indicate the presence of surprising quantities of oxygen-hydrogen (OH) combinations, or *hydroxyl*.

Two-atom combinations were surprising enough, but then three-atom combinations began to turn up. In late 1968 and afterward, radio-wave absorption was found to indicate the presence of such three-atom combinations as water (H_2O) and hydrogen cyanide (HCN) and the four-atom combination of ammonia (NH_3).

Even more complicated molecules were detected: formaldehyde (HCHO), formic acid (HCOOH), and cyanoacetylene (HCCCN). The most complicated ones yet found are the six-atom combinations of methyl alcohol (CH_3OH) and formamide (NH_2COH). The last is the first to be detected containing all four major combining atoms: nitrogen, hydrogen, carbon, and oxygen.

How these complicated molecules are formed, astronomers don't know, but now the necessity arises of studying the chemistry of very thin gas clouds — a science called *astrochemistry*.

Astronaut

MAN HAS ALWAYS DREAMED of flying, of winging through the sky like a large bird. In ancient times, it was felt the air continued indefinitely upward, even to the heavenly bodies. As late as 1638, an English bishop, Francis Godwin, wrote a fanciful tale of a man who was carried to the moon by large birds.

In 1643, an Italian physicist, Evangelista Torricelli, weighed the atmosphere and it became obvious it could only extend a few miles upward. The moon was separated from Earth by over 200,000 miles of vacuum and other heavenly bodies were farther away still.

In 1657, a French poet, Cyrano de Bergerac, in a work published posthumously, first mentioned the possible use of the rocket principle in traveling through space. This is the one practical method of crossing a vacuum and it was by rockets that men actually reached the moon in 1969.

Space is what separates two objects. People are more interested in objects than in the space between them, so space came to be thought of as nothingness. The ideal nothingness is a vacuum and since a vacuum exists beyond the atmosphere and between the heavenly bodies, it is that region which came to be known as *space*, particularly.

In the twentieth century, one could talk of *space flight* for journeys beyond the atmosphere and *space travel* for journeys to the moon or other outer-space destinations. (*Rocket flight* and *rocket travel* might also be used, since rockets were the means of propulsion.)

About 1930, the term *astronautics,* from Greek words meaning "star voyaging," originated in France. It will be a long time before man will be able to journey between the stars to be sure, and for now he must confine himself to the solar system, the neighborhood of his own star. Nevertheless, the word has stuck, and in the United States we call those men who make rocket flights *astronauts* — that is, star voyagers.

ATP

In 1905, two English chemists, Sir Arthur Harden and W. J. Young, were studying the effect of phosphate ions on certain enzymes. To their surprise, the phosphate ions seemed to disappear. A search revealed them to be attached to sugar molecules and these were the first *organic phosphates* to be studied. By the 1920s, it began to appear that organic phosphates were essential to the production of energy in living tissue.

In 1941, the German-American chemist Fritz A. Lipmann found that certain phosphate-containing compounds could release energy in higher than ordinary amounts. These were called *high-energy phosphates.*

Apparently, when food molecules are broken down, phosphate groups are added. At certain key points, chemical changes are brought about that turn the organic phosphate into a high-energy phosphate. Some of the chemical energy in the food is concentrated in that phosphate group, and could be made use of at will.

Of the high-energy phosphates, one in particular was lower in energy than the rest. It stood in an intermediate position. It was capable of accepting a phosphate group from higher-energy phosphates and then passing them on, along with its energy load, to ordinary molecules. Because of this phosphate's use in almost every reaction requiring energy, it was clearly of key importance to the body.

This compound had been discovered in muscle in 1929 by the German chemist K. Lohmann long before the nature of its role was suspected. It was named *adenosine triphosphate* because it consisted of adenosine, a well-known tissue component, to which were attached three phosphate groups in a row. When its importance in energy utilization was made clear, it came to be referred to so often by chemists that a short cut for its heptasyllabic name seemed desirable. The initials *ATP* therefore came into use.

17

Autoradiography

PROBABLY THE MOST IMPORTANT SET of chemical reactions on earth is that of photosynthesis, whereby green plants use the energy of sunlight to convert carbon dioxide and water into their own tissue substances. It is upon plant tissue that all animal life exists.

Scientists have tried to penetrate the details of the photosynthetic reaction. The number of different substances present in plants is, however, enormous and there is no way of telling, by ordinary methods, what was formed first, what next, and so on.

In 1948, the American biochemist Melvin Calvin began a series of experiments designed to work out the details. He developed a scheme for exposing green plant cells to carbon dioxide containing radioactive carbon atoms for just a few seconds, after which the cells were killed so that the photosynthetic reaction was stopped. In the few seconds of exposure, only the first few steps in the reaction had time to take place. If one then analyzed the contents of the cells, and searched only for those compounds that contained radioactive carbon atoms, one could locate the places where the carbon dioxide entered the system.

To do this, Calvin mashed up the cells and dissolved the contents. He then let the solution seep up some porous filter paper. The various substances in the solution seeped upward, each at its own rate. In the end, he had small quantities of particular substances, each concentrated in a certain area of the filter paper.

Calvin then placed a photographic film under the filter paper. Those areas containing substances with radioactive carbon atoms gave off radiations that fogged the film. In this way, he could identify the spots containing compounds to be studied further.

This process of having radioactive compounds indicate their own presence is called *autoradiography*, which means, literally, "own-radiation marking."

Auxin

PLANTS ARE, in general, simpler than animals. Plants, for instance, lack the nerves and muscles which enable animals to move quickly. Yet plants do move, even though only slowly. A plant stem will turn upward and grow in the direction opposite to the pull of gravity; it will also turn to face the light.

That plants can do so is the result of differential growth. Suppose a stem is forced to remain horizontal, along the ground. The cells on the lower side of the stem will grow and multiply more quickly than those on the upper side. The lower side will elongate and the stem will naturally curve upward. In the same way, if a stem is illuminated by light coming from one side only, cells on the shaded side will grow and multiply more quickly and the stem will turn toward the light.

But what stimulates the growth? Growth is stimulated by a hormone, an organic compound produced naturally in plants. The hormone concentrates in those places where growth is required and stimulates that growth. Because these particular hormones are found exclusively in plants, not animals, they are called *plant hormones* or *phytohormones*. (The prefix "phyto-" comes from a Greek word meaning "plant."

The notion of plant hormones was first worked out in detail by a Dutch botanist, Fritz W. Went, in the 1920s. The plant hormones he studied were called *auxins*, from a Greek word meaning "to increase."

Chemists have produced synthetic compounds which have auxinlike properties. These can be used in a variety of ways — to prevent flowering, to keep fruit hanging on the bough for longer periods, or, occasionally, to produce seedless fruits. They are most commonly used as weed killers, however, since in high enough concentrations they seem to overstrain the growth mechanism of plants and, in doing so, kill them.

Avidin

In 1936, two Dutch chemists isolated a substance that in small quantities was essential to life. It proved to be one of the family of vitamins that make up the *B complex* (so called because the first of the group was called *vitamin B* to distinguish it from two others labeled *A* and *C*, the letters being assigned arbitrarily). Because the new vitamin seemed to be found in all forms of life, it was named *biotin*, from the Greek word for "life."

To test its functions, the chemists tried to rear animals on diets lacking biotin, to see what would happen if it were not present. Almost any diet, however, had enough biotin for the needs of the animals.

It had earlier been discovered, though, that rats fed diets that included large quantities of raw egg white, suffered a certain disorder call *egg-white injury*. This was prevented if certain other foods were also added to the diet. By 1940, biochemists showed that it was the biotin in the added food that prevented egg-white injury.

There was something in raw egg white, apparently, that combined with biotin, producing a biotin deficiency that was responsible for the egg-white injury. If enough biotin was added to the diet, this was prevented.

Heating the egg white destroyed its ability to tie up biotin, showing the substance in question to be a protein (for proteins are heat-sensitive). The protein was finally isolated and named *avidin* because it combined so avidly with biotin.

Avidin is not a danger to human health, for few people eat so many raw eggs as to induce abnormal symptoms. Nor is it known what the function of avidin in egg white is. However, avidin has been useful to biochemists who could, with its help, devise biotin-deficient diets through which to study the function of the vitamin.

Barn

PHYSICISTS PRODUCE nuclear reactions by aiming energetic subatomic particles at pieces of matter. If such a particle hits an atomic nucleus, it is likely to alter it somehow. If, on the other hand, it misses the atomic nucleus, nothing happens.

Naturally, physicists can't aim the particles because they are working with objects far too small to see. They can only make use of many particles and hope that some will just happen to score hits.

The number of hits scored by a particular volley of particles depends on the nature of the particles and of the target. Sometimes very few hits are made, as though the target nuclei are very small and therefore easy to miss. Sometimes, under similar conditions, quite a few hits are made — and the nuclei seem bigger in that case.

Physicists would say that nuclei that were easy to hit under certain circumstances had a large *nuclear cross-section;* those that were hard to hit had a small one. Even an easy-to-hit nucleus has a cross-section of only about 10^{-24} square centimeters. This means that a trillion trillion nuclei jammed closely together would cover an area of only one square centimeter.

In 1942, American physicists were working on the atomic bomb and they didn't want to talk too openly about what they were doing. They were bombarding uranium nuclei with neutrons under conditions designed to bring about as many hits as possible. Success was achieved to the point where two young physicists, Marshal G. Holloway and C. P. Baker, said the uranium nuclei seemed "as big as a barn."

For that reason, an area of 10^{-24} square centimeters was called a *barn*. For a while, this was used only out of a desire for secrecy; people overhearing the phrase would not connect it with atomic bombs. Eventually, however, it was accepted as the legtimate unit of nuclear cross-section.

Barnard's Star

As LONG AGO AS 1718, the English astronomer Edmund Halley was able to show that stars actually move. Stars are so distant, however, that it can take many centuries for the motion to become noticeable. The only stars that have motions that are relatively easy to spot are among those that are unusually close to us.

In 1916, the American astronomer Edward E. Barnard noted a dim star in the constellation Ophiuchus (too dim to see without a telescope) moving with a velocity faster than that of any other known star. It moves so quickly that in 180 years it traverses a distance equal to the diameter of the moon. So unusual was this property that the star was one of the few that came to be known by its discoverer's name — Barnard's Star.

The reason for the rapid motion of Barnard's Star is that it is very close to us — only six light-years away. Only Alpha Centauri is closer.

Since 1943, it has been found that the motion of some of the stars nearest to us wavers slightly. The only reasonable cause for that would be the gravitational influence of a large planet revolving about a star. To be detected at a distance of light-years, the wavering had to be considerable, which meant the star had to be smaller than the sun and the planet larger than Jupiter.

In 1963, wavering was detected in Barnard's Star. The Dutch-American astronomer Peter Van de Kamp decided, in 1969, that the wavering might best be explained in terms of two planets, one slightly larger than Jupiter, one slightly smaller. It was the first time any star other than our own sun has been found to show signs of having more than one planet. And the planets of Barnard's Star (if Van de Kamp is correct) are the smallest ever detected outside our solar system.

Baryon

IN THE EARLY 1890s, physicists were studying radiation produced when an electric current was sent across a vacuum. It became clear, eventually, that the radiation was composed of tiny particles, much smaller than atoms. The particles, called *electrons,* carried a negative electric charge. All atoms were found to contain electrons.

But atoms are electrically neutral. If they contain electrons, they must also contain particles with a positive electric charge to balance them. The search was on for such a positively charged particle.

These particles were found and studied in succeeding years, but they were much more massive than electrons. Even the smallest positively charged particle that was found was about 1836 times as massive as the electron. In 1914, the New Zealand–born physicist Ernest Rutherford decided this positively charged particle was responsible for almost all the mass of matter. It was of first importance to the mass and he therefore called it a *proton,* from a Greek word meaning "first." He maintained that the smallest atom, that of hydrogen, was composed of one proton and one electron with the opposite charges exactly balanced.

In 1930, another massive particle, just about as massive as the proton, was discovered by the English physicist Sir James Chadwick. He named it *neutron* because it was electrically neutral.

The fact that the proton and neutron are so much more massive than the electron is very important. The proton and neutron are in the atomic nucleus at the very center of the atom, so that almost all the mass is there. The light electron remains in the atom's outer reaches. The massive particles interact in ways that are fundamentally different from the behavior of the light electrons. For this reason the massive particles, including the proton, the neutron, and still more massive particles since discovered, are lumped together as *baryons,* from a Greek word meaning "heavy" or "massive."

23

Bathypelagic Fauna

PLANT LIFE is found only in the topmost layer of the ocean, for it depends on light, and sunlight can only penetrate some dozens of yards at most into the water. It is only in this *photic zone* (from a Greek word meaning "light") that plants grow.

Animal life penetrates more deeply, for they feed on each other and on debris that sinks downward continually from the photic zone. Scientists first became aware of this in the 1840s when the English naturalist Edward Forbes dredged up a starfish from a depth of a quarter of a mile. Then, in 1860, a telegraph cable lifted from the bottom of the Mediterranean Sea, a mile down, was found encrusted with life forms.

The beginning of the systematic study of underwater life came in 1872, when the exploring vessel *Challenger*, under the British naturalist Charles W. Thomson in a voyage spanning 69,000 miles, made a thorough attempt to dredge up organisms from the depths.

Deep-sea fish were brought to the surface. Some had luminescent patches in their skins, so that those parts glowed in the dark. *Anglerfish* had fleshy growths on their noses that resembled wriggling worms and attracted smaller fish within reach of the angler's gulping mouth. Fish with extensible stomachs could swallow other fish larger than themselves. It was a strange new world that was revealed.

There were deep-sea creatures other than fish, too. There were protozoa, jellyfish, worms, shellfish, and so on. All together these are now referred to as *bathypelagic fauna* (from Greek words meaning "deep-sea animals"). The term is properly applied to animals that float or swim in open water; but there are also animals attached to the deep-sea bottom. These latter are *benthic fauna* from another Greek word referring to the deep sea. The true study of benthic fauna came in the 1960s, when men penetrated to the very deepest part of the ocean and found life on its floor.

Bathyscaphe

ONE DIFFICULTY that stands in the way of man's exploration of the ocean depths is the enormous pressure resulting from the weight of miles of water. A man in even the most elaborate diving suit cannot safely descend more than 300 feet or so.

In the early 1930s, the American naturalist Charles W. Beebe designed a thick-walled spherical steel vessel with thick quartz windows into which a man could fit and be lowered to great depths. Beebe called his device a *bathysphere*. The prefix "bathy-" is from a Greek word meaning "deep," so it was a "ball of the deep."

In 1934, Beebe descended 3000 feet in his bathysphere, while a co-worker, Otis Benton, finally reached a depth of 4500 feet in 1948. Benton used a modified bathysphere he called a *benthoscope* (from Greek words meaning "to see the sea depths").

Neither the bathysphere nor the benthoscope was maneuverable. In 1947, the Swiss physicist Auguste Piccard designed a bathysphere that was suspended from a dirgiblelike bag containing gasoline. Since gasoline is lighter than water, this buoyed up the sphere and let it sink slowly. The vessel carried a ballast of iron pellets, which it could jettison. The vessel would then be lightened and lifted upward by its overhead gasoline bag. The vessel also had electrically driven propellers to allow it to move horizontally. The device was called a *bathyscaphe* ("ship of the deep").

In 1960, the inventor's son, Jacques Piccard, and an American naval officer, Don Walsh, stepped into the *Trieste*, a bathyscaphe built by the Piccards, who sold it to the U.S. Navy. In it, they prepared to explore the Marianas Trench in the western Pacific, the deepest spot anywhere in the ocean. They plumbed downward over seven miles to the very bottom and then returned safely to the surface. The entire sea had been opened to human exploration.

Berylliosis

As TECHNOLOGY DEVELOPS, human beings are exposed to hazards that did not exist in a pretechnological era. For instance, prior to the twentieth century, there was very little opportunity for human beings to encounter concentrated sources of radiation such as x rays or gamma rays. Nor was there any chance at all of encountering various synthetic organic compounds, which never existed in nature but which were produced in the chemical laboratory — some of which proved instrumental in causing cancer.

Even substances which do occur in nature may exist only in small concentrations or in relatively harmless form but be concentrated and made dangerous by the requirements of technology.

Thus, once fluorescent lights were developed in the mid 1930s, there was need for *phosphors* (from a Greek word meaning "to carry light"), which give off visible light after absorbing ultraviolet light. These phosphors were used to coat the inner surfaces of the fluorescent tubes and one of them was a powdered compound of the light metal beryllium.

In 1946, an increasing number of illnesses were noticed among the workers engaged in the manufacture of fluorescent lights. The symptoms, involving the lungs, sometimes appeared months or even years after the patient had ceased working on fluorescent lights, and the illness usually ended in death. The blame was soon pinpointed on the beryllium compound used and it was dropped from the list of phosphors.

This relieved a serious danger. Not only those working in the field ran risks, but also anyone who accidentally broke a fluorescent light, breathed its dust, or was cut by the jagged glass.

The disease, now no longer an acute threat, is called *berylliosis*, the suffix "-osis" being commonly used in medicine to signify a pathological state.

Big Bang

How DID THE UNIVERSE come into being? A hint at a plausible answer came in the 1920s when the American astronomer Edwin P. Hubble showed that the distant galaxies were receding from us in a very systematic way, as though the entire universe were expanding.

In 1927, the Belgian mathematician George E. Lemaître pointed out that if we looked upon the movement of the universe in reverse, as though we were reversing a movie film, it would seem to be contracting — eventually into a hard mass. Lemaître suggested that all the mass of the universe was, in the beginning, squeezed into one relatively small, enormously dense "cosmic egg," which exploded and gave birth to the universe as we know it.

Fragments of the original sphere of matter formed galaxies, still rushing out in all directions as a result of that unimaginably powerful multibillion-year-old explosion.

The Russian-American physicist George Gamow went on to elaborate this notion. He calculated the temperatures in the fragments of that explosion; how quickly each temperature would drop; how the initial energy would be converted into subatomic particles, then simple atoms, then complicated ones. Thinking of that first tremendous explosion, Gamow called it the *big bang* theory of the universe's origin and the name has stuck.

The big bang theory was just theory to begin with. How to gain evidence for it? In 1965, two American radio astronomers, Arno A. Penzias and R. W. Wilson, showed that there was a general background of radio waves from every part of the sky. This seemed to be the remnant of the radiation of that vast explosion, still lingering after all those billions of years. This is, at the moment, the strongest piece of evidence in favor of the big bang.

27

Binary Digit

OUR USUAL SYSTEM for expressing numbers makes use of powers of 10. In the number 222, for instance, the first 2 is two hundred (2×10^2), the second 2 is twenty (2×10^1) and the third 2 is two (2×10^0); so 222 is really $(2 \times 10^2) + (2 \times 10^1) + (2 \times 10^0)$. It is important to remember that 10^0 or any other number to the zeroth power equals 1.

Any number other than 10 can be used as a base. In a five-based system, 222 would be $(2 \times 5^2) + (2 \times 5^1) + (2 \times 5^0)$, or 62 in the ten-based system. In a three-based system 222 would be $(2 \times 3^2) + (2 \times 3^1) + (2 \times 3^0)$, or 26 in a ten-based system.

In any system, the number of different digits, counting zero, is equal to the value of the base. There are ten digits in our ordinary system, but only five in a five-based system: 0, 1, 2, 3, 4. The digit 5, in the five-based system would be expressed 10, our 6 would be 11, and so on. In the three-based system, only 0, 1, and 2 would exist, and so on.

The simplest system is the two-based system, which would have as digits only 0 and 1. In the two-based system, a number like 1101 would be $(1 \times 2^3) + (1 \times 2^2) + (0 \times 2^1) + (1 \times 2^0)$, or 13 in the ten-based system. Writing numbers in order in the two-based system, we would have 1, 10, 11, 100, 101, 110, 111, 1000, 1001, 1010, and so on.

Computing in the two-based system is very simple, especially for electronic computers. Every on-off switch within a computer can represent 0 in the off position and 1 in the on position. The switches can be designed to maneuver according to two-based arithmetic and give answers with the speed of the electric current. The two-based system is also called the *binary* system from a Latin word meaning "two at a time." Therefore, 0 and 1 are *binary digits*, which can be abbreviated *bits*. Zero and 1 are the smallest pieces of information that can be fed a computer, so that we can speak of so many "bits" of information.

Bioastronautics

As soon as man succeeded in launching satellites into orbit about Earth, the question naturally arose: how would man fare in space? If space was to be explored, men would have to be sent beyond the atmosphere and to other worlds. Could this be done?

In some ways, the experience of space travel would introduce entirely new stresses on living organisms. There was the high acceleration, for instance, that would have to be endured while the spacecraft was being placed into orbit. Then, there was the experience of weightlessness once the orbit was established. And what of the charged particles and energetic radiation that are present in space — particles and radiation that are ordinarily filtered out by our atmosphere and never reach organisms on Earth's surface?

To go further, there are questions as to how a human being can be kept alive inside the space capsule even when there are no outside dangers. How is the air to be kept pure; how is the occupant to be fed; how are wastes to be disposed of?

All this can be, and has been, called *space biology* or *space medicine*, but a more formal name for the study is *bioastronautics* (from Greek words meaning "life aspects of space travel").

The science began on Earth, for the effect of acceleration can be studied on life forms by whirling them rapidly in centrifuges. It entered space itself early, however. The second satellite to be placed in orbit (*Sputnik II*, launched by the Soviet Union in November 1957) carried a dog, whose reactions were monitored by instruments.

Since then, many men have been placed in orbit; some have stayed in space for weeks without harm; some have reached the moon and returned; and all seem well. The same is true of the one woman launched in space (by the Soviet Union). In fact she married a man who had been in space and they have a seemingly normal child.

29

Bionics

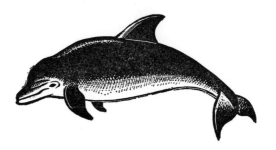

MANY OF THE EFFECTS man is trying to achieve by means of his machinery have been obtained in living systems during billions of years of trial and error. The devices of life, however, are sometimes not suitable for machines. Much time, for instance, was wasted by inventors who imitated birds and tried to design flying machines with flapping wings.

On the other hand, consider dolphin skin. Dolphins swim at speed that would require 2.6 horsepower if the water about them were as turbulent as it would be about a vessel of the same size. Water flows past the dolphin without turbulence, however, because of the nature of dolphin skin. If we could imitate the effect in ship walls, the speed of an ocean liner could be increased while its fuel consumption was decreased.

Again, the American biophysicist Jerome Lettvin studied the frog's retina in detail by inserting tiny platinum electrodes into its optic nerve. It turned out that the retina had five different types of cells, which reacted (1) to sudden changes in illumination from point to point, (2) to dark, curved surfaces, (3) to rapid motion, (4) to dimming light, (5) to the blue of water. All combined to make seeing extraordinarily efficient for the frog. If man-made sensors could be made to use such tricks they would become far more versatile than they now are.

In short, engineers are now making conscious efforts to adapt biological systems to man-made electronic devices. In 1960, an American engineer, Jack Steele, shortened the descriptive term *biological-electronics* to *bionics*.

In a sense, the ultimate goal of bionics is to invent a device that would imitate the functioning of the human brain and give us that science fictional dream, the robot.

Biosphere

THE ANCIENT GREEKS considered the universe to be made up of a series of concentric shells. At the center was the solid ball of Earth itself, which we now call the *lithosphere* (from Greek words meaning "stone ball"). Surrounding that, though not quite entirely, was a shell of water, the *hydrosphere* (water ball), and outside that, the *atmosphere* (vapor ball). Beyond the atmosphere the Greeks envisaged a sphere of fire and various planetary spheres, theories which we have abandoned. However, we have divided the atmosphere into various regions, using the same suffix — *troposphere*, *stratosphere*, and so on.

In one respect, though, modern scientists have gone ahead and added another shell. In a thin shell around Earth, including the hydrosphere, the outermost part of the lithosphere, and the lower reaches of the atmosphere, are included all the life forms of the planet and all the living activity. This was first pointed out by the French zoologist Jean Baptiste de Lamarck at the beginning of the nineteenth century, and the concept was sharpened by the Austrian geologist Eduard Suess in 1875. This thin shell in which all of life lives is the *biosphere* (life ball).

The total mass of living things on Earth is estimated to be about seventeen billion billion tons. This is only about 1/300 the mass of the atmosphere, and only 1/70,000 the mass of the ocean. Nevertheless, the biosphere is so active chemically that it is responsible for much of the environment about us. For instance, it is thanks to the green plants that the air is full of free oxygen. Also, the great layers of limestone produced by corals make up large atolls and long reefs.

On the other hand, the biosphere is extraordinarily fragile and even small changes in temperature or the addition of small quantities of poisons could change it drastically or wipe it out altogether. Man himself has been putting great strains on the biosphere in recent years and this is of growing concern to scientists.

31

Black Hole

In 1916, Albert Einstein, in his *general theory of relativity*, suggested a new way of looking at gravitation. He said that masses distorted space in such a way that objects found the shortest distance between two points to lie along a curved path. Objects passing by each other therefore followed this curved path. If they were close enough to each other, the curvature would be so sharp that they would follow closed paths about each other.

Einstein differed from the older suggestions of Sir Isaac Newton in his belief that, in the relativistic view of the universe, light rays also followed a curved path in the presence of masses. Because the photons that made up the light moved so much more rapidly than ordinary astronomical bodies, they had a chance to curve only very slightly before moving on far beyond the space-distorting mass.

Still, the amount by which the path of light curves depends on the mount of mass present and its concentration. If a very large quantity of mass is concentrated into a very small volume, then space is enormously distorted in its immediate vicinity. Light passing near such a body would be trapped into a closed orbit and would never escape. This was pointed out in 1916, almost as soon as Einstein's paper appeared, by a German astronomer, Karl Schwarzschild.

In the vicinity of this massive compressed body, not only would light be trapped but so would everything else. It would be like a hole in space into which things could fall, but from which nothing, not even light, could emerge. It would be, therefore, a *black hole*.

For many years this was considered only theory, but now the newly discovered pulsars seem to be neutron stars. Neutron stars are almost massive enough and compressed enough to be black holes, and so astronomers are now searching avidly for any sign of the existence of the thing itself.

Borane

THE SIMPLEST HYDROCARBON (a substance made up of carbon and hydrogen atoms only) was named *methane*. As a result, the "-ane" suffix was applied to many hydrocarbons and then to compounds of hydrogen with one other element, even when that element was not carbon.

At the time this convention was established, there were already compounds of hydrogen that had names of other sorts that could not be abandoned. Two hydrogen atoms with an oxygen were *water*, three hydrogens with a nitrogen were *ammonia*, three hydrogens with phosphorus, or with arsenic, or with antimony, were *phosphine*, *arsine*, and *stibine*, respectively. (The prefix "stib-" is from "stibium," the Latin name for an antimony compound used in ancient times.)

Silicon, which closely resembles carbon, was found later to form compounds with hydrogen and these were called *silanes*. Like carbon, silicon atoms can form chains, so that we can have one silicon atom combined with four hydrogen atoms (monosilane), two connected silicon atoms combined with six hydrogen atoms (disilane), and so on.

Germanium, tin, and lead (in the same family as carbon and silicon) also form compounds with hydrogen, which are *germane*, *stannane*, and *plumbane*, respectively. (The prefixes "stann-" and "plumb-" are from *stannum* and *plumbum*, Latin for "tin" and "lead.")

One other element is of importance in this respect. Boron combines with hydrogen atoms to form *boranes*. A single boron atom should combine with three hydrogen atoms to form *monoborane* or simply *borane* but this has not been isolated. Chemists have, however, obtained *diborane*, with two boron atoms and six hydrogen atoms. Higher analogs with molecules containing up to twelve boron atoms (dodecaborane) are also formed. Boranes became important in the 1950s when they were added to rocket fuels to give them greater thrust.

Borazon

THE HARDEST SUBSTANCE KNOWN is diamond, which is made up exclusively of carbon atoms. The small carbon atoms can come unusually close to each other, so the attraction between them is great. Each carbon atom has four electrons in its outermost shell, which can hold eight. Each therefore shares two electrons with each of four neighbors, so that each has a share in eight electrons altogether. No other naturally occurring substance can form so many strong bonds in so many different symmetrically placed directions. (Carbon atoms can form bonds in an asymmetric pattern, as in graphite or coal, and the substance is then not particularly hard.)

A boron atom has one electron less than carbon, and has only three in its outermost shell. A nitrogen atom, with one electron more than carbon, has five. Both are small atoms, and when boron, in elemental form, has each atom bonded in three directions, it is pretty hard, though certainly not as hard as diamond. Nitrogen atoms combine in pairs and no more, so that the element is gaseous.

But imagine equal numbers of boron and nitrogen atoms bonded in alternation. A boron and nitrogen atom, together, would have a total of eight electrons in the outermost shells, just as a pair of carbon atoms would, and there should be similarities. Boron nitride, with molecules consisting of an atom of each element (BN), is indeed something like graphite in its properties. Graphite, under huge pressures, shifts to a symmetrical atomic arrangement and becomes diamond. What of boron nitride? At a pressure of 65,000 atmospheres and a temperature of 1500° C. (a combination not available till the 1950s), the symmetrical form is produced — a kind of boron nitride as hard as diamond. It is called *borazon* (the second part of the name from *azote*, an old name for "nitrogen"). Borazon has the advantage over diamond of being much more resistant to heat.

Botulism

A VERY COMMON BACTERIUM is *Clostridium botulinum*. *Clostridium* is Latin for "little spindle," which describes the bacterium's shape and *botulinum* is from the Latin word for "sausage," where it is sometimes detected.

C. botulinum is *anaerobic* (from Greek words meaning "no air") since it can only live in the absence of oxygen. Under unfavorable living conditions, it, like many other bacteria, forms a hard pellicle about itself and becomes a spore. Within the spore the fires of life sink low and it can survive extreme conditions without actually dying. It remains ready to become actively alive again when conditions improve.

In canning food, or in making preserves, the product must be boiled after sealing. To make sure that the hardy spores are killed, the boiling must be continued for at least half an hour. Ordinary bacteria cannot live inside the can, where oxygen is lacking, but C. botulinum can, for it requires no oxygen. If its spores survive, they become active, grow, and liberate *botulinum toxin* which is the most poisonous substance known. (An ounce of the toxin, properly distributed, would be enough to kill every human being on Earth.)

If food containing the toxin is eaten, the toxin is very slowly absorbed and affects certain nerves leading to muscles. There is a selective muscle paralysis affecting the eyes and throat first so that those suffering from *botulism* find it difficult to focus their eyes and to speak. Then the chest muscles are paralyzed and it is the inability to breathe that kills.

Every once in a while an imperfectly heated batch of cans will lead to cases of botulism. Even one case is sufficient to set in motion a thorough search for any other possibly infected cans, so dreaded is the disease. One such scare, which received a great deal of publicity, took place in the summer of 1971, for instance.

Bremsstrahlung

IN 1895, the German physicist Wilhelm C. Roentgen discovered a new kind of radiation so energetic it could pass through glass or cardboard. He did not know its nature so he called it *x rays*. The radiation turned out to be like ultraviolet radiation but with smaller and more energetic waves.

Roentgen had obtained his x rays through the action of cathode rays (speeding electrons) as they collided with the glass of the tube within which they were produced. What if the cathode rays were allowed to strike a piece of metal instead? Would the situation change? Metal pieces were inserted into the cathode ray tube and when the stream of electrons struck, x rays were produced in greater quantity and possessing greater energy.

It was found by the English physicist Charles G. Barkla, in 1911, that the energy content of the x rays depended on the nature of the metal used to stop the electrons. The English physicist Henry G.-J. Moseley went on to deduce from those energies the structure of the atomic nuclei of the various metals. From this, in 1914, he proceeded to work out the concept of the *atomic number*, which brought final order to the list of elements.

But how do the x rays originate? The speeding electrons have a great deal of energy of motion. When they collide with some substance with which they interact, they decelerate rapidly and lose this energy.

Energy, however, cannot be destroyed. If the electrons lose energy of motion, this energy must reappear in some other form. As it happens, it appears in the form of radiation. The Germans, who first studied this, called the radiation *Bremsstrahlung* (deceleration radiation) and the phrase was accepted by English-speaking physicists without translation. X rays, then, are a form of bremsstrahlung.

Brownian Motion

In 1827, the Scottish botanist Robert Brown was viewing a suspension of pollen grains under a microscope. He noted that the individual grains were moving about irregularly even though the water in which they were suspended was quiet. Since the pollen grains had the potentiality of life within them, Brown thought at first the motion was a manifestation of this life. Then, however, he viewed a suspension of dye particles in water. These particles were definitely nonliving and yet they moved about randomly just as the pollen grains had done.

Brown had no explanation for it but, because he reported it, the phenomenon has been known as *Brownian motion* ever since.

In the 1860s, the Scottish physicist J. Clerk Maxwell worked out the rules governing the behavior of gases on the assumption that they consisted of small particles (atoms or molecules) in random motion. The analysis was very convincing and it seemed that liquids, too, would consist of particles in random motion. A grain of pollen or a piece of dye would be bombarded on all sides by randomly moving water molecules and since the force might be slightly greater from one side than another, the suspended grains would be moved first this way, then that.

The amount by which suspended grains of a certain size would move this way and that would depend, in part, on the size of the bombarding water molecules. In 1905, Albert Einstein worked out an equation describing this; and in 1908, the French physicist Jean B. Perrin set about making the necessary observations to determine the size of the values in Einstein's equation.

He succeeded and was able to calculate the size of the water molecules. This was the first direct determination of molecular size and was the final proof that atoms and molecules did indeed exist and were not simply convenient chemical fictions.

Bubble Chamber

UNTIL WELL AFTER WORLD WAR II, the best way of following the tracks of subatomic particles was to use the cloud chamber invented by Charles T. R. Wilson in 1911. It contained air saturated with water vapor. Particles, speeding through, chipped electrons away from the atoms they hit, forming electrically charged ions. Around these ions tiny water droplets formed, making a visible track from which much could be determined concerning the particles.

But air contains relatively few atoms. Particles scored hits only occasionally and the track was therefore thin and light. Events that took place very quickly were easily missed and the fine details of even comparatively slow events weren't clear.

Liquids are denser than gases. A speeding particle would strike many more atoms in a liquid than in a gas. But how to use a liquid? In 1952, the American physicist Donald A. Glaser, while talking over a glass of beer, began to watch the bubbles forming in the beer. It occurred to him that instead of following water droplets in air, you could as easily follow gas bubbles in liquid.

Suppose you heated a liquid to the point where it was about to boil and put it under just enough pressure to keep it from boiling. A subatomic particle, speeding through the liquid, would produce ions, about each of which a tiny bit of boiling would take place. For an instant there would be a visible wake of bubbles left by the particles and this could be photographed. Before 1952 was over, Glaser had built the first *bubble chamber*.

He used ether as the liquid at first, then switched to liquid hydrogen at extremely low temperatures. At low temperatures, efficiency was greater and the tracks clearer. With bubble chambers, very rapid particle interactions could be observed in great detail.

Calcitonin

A HORMONE generally has a powerful effect on some aspect of body chemistry, even when it is present in only tiny quantities. The presence of *insulin*, for instance, causes the blood content of glucose to go down. Since there is a desirable level for blood glucose, with too low a level being as bad as one too high, one doesn't want insulin to be present in too great a quantity. Feedback is therefore used. If the level of blood glucose drops, insulin production is inhibited. As the insulin vanishes, the level of blood glucose rises, and insulin is produced again.

Feedback does best when there are two hormones, working in opposite directions. Thus a second hormone, called *glucagon*, is also formed by the same gland that forms insulin. Glucagon's action is the reverse of insulin's.

Such double action from opposite directions was searched for elsewhere. The parathyroid glands produce a hormone (called *parathormone* after its organ of origin) which raises the level of calcium ions in the blood. Was there an opposition hormone to that?

In 1963, the Canadian physiologist D. Harold Copp worked with glands from 30,000 pigs and managed to extract about a tenth of a gram of a substance that did indeed have an effect opposed to that of parathormone. It lowered the level of calcium ions in the blood. Copp named it *calcitonin* because it participated in regulating the *tone* (that is, the concentration) of calcium in the blood.

At first, it was assumed that the calcitonin was formed in the same gland as its opposite number (parathormone), just as was the case for those other two opposites, insulin and glucagon. This proved not to be so. By 1967, it was clear that calcitonin was formed not in the parathyroid gland, but in the nearby thyroid gland. To emphasize this, the hormone is sometimes called *thyrocalcitonin*.

Carbonaceous Chondrites

UNTIL ROCKS AND SOIL were brought back from the moon in 1969, meteorites were the only matter known that was not originally part of Earth.

Meteorites can be classified by their chemical structure. Some, for instance, are almost purely metallic, made up of a mixture of iron and nickel in proportions of about 9 to 1. Others are stony in nature, made up of minerals very like those making up the deeper sections of Earth's rocky portion.

There are, however, other ways of classifying the meteorites. Of the stony meteorites, over 90 per cent include small, compact spheres of substance within them. These spheres are called *chondrules*, from a Greek word meaning "grain of wheat," because they are frequently about the size of such grains. Meteorites containing these inclusions are called *chondrites;* those that don't are called *achondrites.* (The Greek prefix "a-" means "not.")

The chondrites aren't ordinarily very different from the material surrounding them. They, too, are rocky in nature and resemble the kind of substances in Earth rocks.

A small proportion of the chondrites, however, are black in color, and the chondrules they contain have a considerable percentage of carbon. These are called *carbonaceous chondrites* in consequence, and they are by far the most fascinating of all the meteorites since carbon is the basic element of life.

Can the carbonaceous chondrites represent traces of life stemming from another planet? In 1961, some American scientists actually reported finding shapes in the chondrules that might once have been part of life forms. The furor soon died down, however, as it turned out the suspected shapes were contamination from the earth around them. Nevertheless, the presence of carbon itself remains intriguing.

Carborane

THE BORANES ARE COMPOUNDS of boron atoms and hydrogen atoms which came to be of increased interest in the 1950s because they were used as rocket fuel additives to give more thrust. As the boranes were investigated, the boron atoms in their molecules were found to be arranged in geometric patterns — in octahedrons and icosahedrons. An octahedron is a solid with eight faces and six vertices. The six boron atoms in *hexaborane* would be distributed, one at each vertex. An icosahedron has twenty faces and twelve vertices. The twelve boron atoms of *dodecaborane* are distributed one at each vertex.

The carbon atom is very like the boron atom in size, though it contains one additional electron. In 1963, it was found that one or more carbon atoms could substitute for boron atoms in these geometric patterns. This meant the discovery of an entirely new class of materials in which hydrogen atoms are bound to an interconnected web of boron and carbon atoms. These are the *carboranes*.

Even when most of the boron atoms are substituted by carbon atoms, the borane structure is retained. The presence of the carbon atoms, however, gives the molecule much greater stability. Indeed, the more carbon atoms present, the more stable the compound is to heat and the more inert to chemical reaction. There are uncounted numbers of possible carboranes since one or more of the hydrogen atoms can be replaced by other atoms or groups of atoms.

As yet the compounds exist only in small quantities and are merely laboratory curiosities, but as knowledge concerning them increases, and as better methods of production are worked out, they will be formed in larger quantities. We may expect to find uses for them that will take advantage of their particular properties so that whole new groups of polymers and plastics may become available.

41

Cepheid

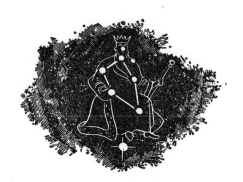

IN THE GREEK MYTHS, Cepheus was a king of Ethiopia whose daughter, Andromeda, was saved from a sea monster by the hero Perseus. The Greeks pictured the stars in constellations representing mythical characters and one of these was labeled *Cepheus*. The brighter stars in each constellation are named in order of brightness by Greek letters. The fourth letter of the Greek alphabet is *delta*, so the fourth brightest star of the constellation Cepheus is *Delta Cephei*.

Delta Cephei was found to be an unusual star in that its brightness varied regularly (it is a *variable star*). The star brightens and dims in a fixed pattern every 5.37 days. Eventually, other stars were found that brightened and dimmed in similar fashion, but with periods of anywhere from two to forty-five days. All variable stars with this sort of pattern were lumped together as *Cepheids*.

In 1912, Cepheids took on an unexpected importance, when the American astronomer Henrietta S. Leavitt was able to show that the period of variation was closely related to the amount of light given off by the star (its *luminosity*).

This meant that from its period alone, one could tell how luminous a Cepheid was. Then, from its apparent brightness, one could tell how far away it must be. It was by studying Cepheids that the American astronomer Harlow Shapley was first able to show the size of our galaxy — to demonstrate that it had to be 100,000 light-years across. He was also able to show that the solar system was nowhere near the center of the Galaxy (as had been earlier thought) but was far out to one end.

By using the Cepheid yardstick, it also became possible, for the first time, to measure the distance of some objects beyond the Galaxy. The Magellanic clouds, for instance, were found to be about 150,000 light-years from us.

Čerenkov Radiation

THE SPEED OF LIGHT in a vacuum (186,282 miles per second) is considered the limit of possible velocity for any particle possessing mass. However, when light passes through some medium other than a vacuum, it moves more slowly. Through water, light travels at a velocity of only 140,000 miles per second; through glass, only 110,000 miles per second; and through diamond, only 77,000 miles per second.

It is possible for subatomic particles, as they move along at nearly the speed of light in a vacuum, to smash through water or glass at nearly their speed in air. Their speed in water or glass is therefore far greater than the speed of light *in water or glass*.

As the particles race through the medium, they slow up somewhat and the energy they lose appears as radiation. A bluish light is emitted, but it cannot keep up with the speeding particles. It trails off behind like the wake behind a motorboat, and the angle it makes with the line of motion of the particle depends on how much faster than the speed of light the particle is going.

The first to observe this blue light emitted by such very fast particles was a Soviet physicist named Pavel A. Čerenkov, who reported it in 1934. The light is therefore called *Čerenkov radiation*. In 1937, two other Soviet physicists, Ilya M. Frank and Igor Y. Tamm, explained the existence of this light by pointing out the faster-than-light aspect of the particle's motion.

Special instruments, *Čerenkov counters*, have been designed to detect such radiation, measure its intensity and the direction in which it is given off. These are particularly useful in studying very energetic cosmic ray particles which move within fifty miles per second of the speed of light in a vacuum. Such particles move faster than light does in air so that they produce Čerenkov radiation even in air.

Cermet

AMONG THE FIRST MATERIALS used by mankind were certain earthy substances such as clay. If this was baked hard in a fire, the result was something which was hard, insoluble in water, and resistant to fire. Clay could be used to make pottery and bricks.

The Greeks called such baked clay *keramos*. As a result, substances such as clay, which are hard, insoluble in water, and resistant to heat are called *ceramics*.

Most ceramics are oxides, combinations of other elements with oxygen. Thus we may have silicon dioxide, aluminum oxide, chromium oxide, magnesium oxide, and so on. Silicon dioxide (the most common) is often found in combination with one or more of the others, forming what are called *silicates*. Clay is an aluminum silicate. Glass, the various glazes, and porcelain are examples of other silicates. (Glass is not sufficiently resistant to heat to be a good ceramic, but it has the virtue of being transparent.)

The greatest shortcoming of ceramics, generally, is that they are brittle. They will not bend without cracking and a sharp blow will break them altogether. Once metals were discovered, therefore, they replaced stone and ceramics where toughness was needed. Metal knives could hold a sharper edge than flint could and wouldn't blunt as quickly with use. Metal pots could be dropped without breaking. Metal objects could be bent and shaped in a variety of ways.

However, metal would rust, be damaged by water, and be more affected by heat than ceramics. In recent years, techniques have been devised to mix ceramic and metallic powders and compact them together under heat to form a kind of combination substance with a combination of virtues, as heat resistant as ceramics and as tough as metals. This combination is a ceramic-metal or, in abbreviated form, a *cermet.*

44

Chloroplast

THE GREEN OF GREEN PLANTS is due to *chlorophyll* (from Greek words meaning "green leaf"). By way of chlorophyll, the energy of sunlight is used to turn carbon dioxide and water into plant tissue and oxygen. Animal life, including ourselves, depends on those tissues and oxygen to eat and breathe.

In 1865, the German botanist Julius von Sachs showed chlorophyll in plant cells to be confined to small bodies in the cytoplasm. These small bodies were eventually named *chloroplasts*, the suffix coming from a Greek word referring to something with a definite form.

The interior of chloroplasts is divided by many thin membranes called *lamellae*, from a Latin word meaning "little plates." The lamellae thicken and darken in places to form *grana* (from a Latin word meaning "grains"), and within each of these are 250 to 300 chlorophyll molecules.

Some chloroplasts are quite large, so large that there is only one to a cell. There are indications that they possess DNA, something characteristic of cell nuclei. Could it be that chloroplasts were once, in the dim past, independent organisms?

That this may be so is indicated by the fact that the very simplest plant cells do not have distinct chloroplasts or nuclei. The chlorophyll systems and the nuclear material are distributed throughout the cell. Such cells are called *blue green algae*, where *algae* is the general name for unicellular plants living in water (from a Latin word for "seaweed"). Not all blue green algae are blue green in color, but the first ones studied were.

Blue green algae might almost be considered free-living chloroplasts. Bacteria very much resemble blue green algae except for the lack of chlorophyll. Could these two classes of organisms represent the kind of life on Earth before true cells with separate nuclei evolved?

45

Circadian Rhythm

ALTHOUGH WESTERN MAN is wedded to clocks and is constantly aware of the importance of knowing the time, he isn't entirely dependent on them. Even without clocks, he would know it was mealtime when he got hungry, and bedtime when he got sleepy. There are many people who can wake up at some desired time without having to set an alarm.

There are cyclic changes inside you that make you feel hungry or sleepy every so often and that keep you roughly aware of the passing of time. Such cycles are examples of *biological clocks*.

What set the biological clocks? There are steady cycles in the world outside the organism: light and dark alternate; the tides move in and out; the seasons bring rain and drought, or warmth and cold.

It is useful for an organism to respond to these changes. If its food is to be found only by night or only in the warm season, it might as well go to sleep during the day or hibernate during the winter. If it is going to lay its eggs on the shore, it can do so best at the highest high tide that comes with the full moon. Even plants respond to these rhythms so that leaves curl at sunset, flowers or fruits come at particular seasons, and so on.

Undoubtedly, this happens because of basic molecular changes within organisms developed by evolution, and in recent years biologists have been studying the details of these inner "clocks" with great interest.

The strongest rhythm is, of course, that of day and night, with its alternation of warm light and cool dark. There are many cycles that vary over a period of about a day. The Latin phrase for "about a day" is *circa dies*, so the daily rise and fall is called *circadian rhythm*.

These circadian rhythms are hard to ignore. People who make long jet trips find their circadian rhythm to be out of tune with the new position of the sun and it can be very troublesome to try to adjust.

Cistron

THE PHYSICAL CHARACTERISTICS of an individual organism are determined by the genes it possesses. These genes are DNA molecules in the chromosomes within the nucleus of each cell.

Genes have enormous possibilities of variation among themselves because of their complicated molecular structure. A gene governing a particular characteristic always exists in a particular section of a chromosome, but it may exist there in any of a number of varieties. The gene governing eye color may be of a type that produces brown eyes, or one that produces blue eyes. Such varieties of a particular gene are called *alleles* from a Greek word meaning "another."

Chromosomes occur in pairs, so that a gene governing a particular characteristic will be present in each of two chromosomes. The gene may be the same allele in each of the two, or different alleles.

A chromosome can intertwine with its pair in such a way that portions are interchanged (a "crossing over"). The crossover may take place at any point between any two neighboring genes. You might have two genes, a and b, next to each other in one chromosome and two alleles of those genes, a' and b', in its pair. Crossing-over may take place between a and b. As a result the first chromosome may now have a and b' as neighbors while the second has a' and b.

When a and b are in the same chromosome, that is called the *cis* configuration, from a Latin word meaning "on the same side." With a and b on opposite chromosomes, we speak of the *trans* configuration from a Latin word meaning "across."

The a and b genes sometimes act as a unit in producing a particular physical characteristic, even though they can be separated by crossing over. Such an action unit of separable genes is called a *cistron* because they must be in the cis configuration to act normally.

Clock Paradox

ALBERT EINSTEIN's *special theory of relativity*, published in 1905, showed that the measurements of mass and length weren't absolute, but depended on the velocity of the object being measured relative to the instrument doing the measurement.

An object in motion relative to an observer is a little more massive and a little shorter than that same object at rest. At ordinary velocities the change is extremely tiny, but at velocities comparable to that of light in a vacuum the changes mount up. An object moving 162,000 miles a second would be measured as half as long and twice as massive as it would be if at rest. If two objects, *A* and *B*, equal in length and mass move past each other at a velocity of 162,000 miles a second (each relative to the other), *A* would measure *B* as half as long and twice as massive as itself. *B* would measure *A* as half as long and twice as massive as itself also. When the two objects finish their trip and come together again, each would measure the other as normally long and massive, for they are now at rest with respect to each other.

Einstein said this was also true of time. Time on a moving object passes more slowly than on an object at rest. If *A* and *B* are moving at 162,000 miles a second relative to each other, each would observe a clock on the other to be moving at a half-normal rate.

But changing time leaves a permanent mark. When *A* and *B* come together after their trip, *A* would expect *B*'s clock to be slow and *B* would expect the same of *A*'s clock. Rigorous logic appears to bring us to a contradiction here, and that is what we call a *paradox*. Since in the usual account, scientists talk of observing clocks, this particular paradox is the *clock paradox*.

Fortunately, by taking acceleration into account, the clock paradox can be explained away, leaving the theory of relativity intact.

Clone

PLANTS ARE MORE VERSATILE than animals in some ways. A twig from one tree can be grafted to a plant of a completely different variety and that twig will continue to grow and flourish. Or a twig can be planted and develop into a whole plant.

The Greek word for "twig" is *klon* or, in Latin spelling, *clone*. The word has come to refer to any group of cells grown from a single body cell — or to any whole organism developed from a part.

All animals have the power to regenerate lost parts of the body to some extent. Some simple animals can easily regrow entire limbs. Complex animals are less talented, but we can regrow hair and fingernails. We can patch damage to skin or liver, but neither we nor any other mammal can regrow a limb if one is lost.

And yet, somehow it would seem that the cells of the body ought to be able to do more than they do. The fertilized ovum is a single cell that has the capacity to divide and grow into an entire complicated organism. In the process trillions of cells develop from that original cell and those trillions divide into dozens of varieties, none of which can by themselves develop into an organism. In some cases, they can no longer even divide. Yet all the body cells contain the same genes present in the original fertilized ovum.

In each different type of cell, different combinations of genes are blocked and put out of action; the pattern of those that remain determines the cell specialization. But what if the blocked genes are unblocked? One way of doing this might be to expose the cell nucleus to the cytoplasm of a fertilized ovum. In the late 1960s, the nucleus of a frog's fertilized ovum was replaced with nucleus from a frog's body cell. The new cell developed normally and a full-grown frog genetically identical to the frog of the body cell was formed. It was an animal clone.

Cloud Chamber

WHEN SUBATOMIC PARTICLES smash through the atmosphere, they chip electrons away from the atoms they encounter. The electron-missing atoms (called *ions*) can carry away electric charge from metal foil and through this action the presence of subatomic particles and their approximate quantity can be determined. Another method is to have the subatomic particles strike certain chemicals, producing tiny scintillations of light which can be viewed under magnification.

In 1895, a Scottish physicist, Charles T. R. Wilson, was interested in cloud formation. He came to the conclusion that water vapor turned into the tiny droplets that made up clouds through condensation around appropriate nuclei — dust particles or ions of some suitable type. Since subatomic particles formed ions, it occurred to him that water droplets might appear in their tracks.

In 1911, he pumped water vapor into dust-free air in a small closed chamber fitted with a piston. He pulled the piston upward, expanding and cooling the air. Cold air can't hold as much water vapor as warm air, but nothing happened. If there had been dust particles or ions to serve as nuclei, however, water droplets would have formed a tiny cloud.

Repeating the experiment, Wilson sent subatomic particles through the chamber and then expanded the air. This time, a trail of water droplets formed about the ions, making the track of the particle visible. In the presence of a magnet, the subatomic particle's path curved in a direction that depended on whether its charge was positive or negative and by an amount that depended on its mass. The track of water droplets curved too, and from that a skilled observer could tell a great deal about the subatomic particle. For obvious reasons, this detecting device was named a *cloud chamber*.

 # Cloud Seeding

CLOUDS ARE COLLECTIONS of tiny water droplets or tiny ice crystals. Sometimes these collect into conglomerates large enough to fall as rain or snow. At other times, the tiny fragments, too small to fall against air resistance, remain as they are and there is no precipitation. When there is drought, the presence of clouds which do not release their load of moisture is most frustrating.

One way of encouraging the necessary large conglomerates to form is by providing some sort of nuclei about which they can gather. The nuclei can be dust particles of the proper size and shape, or tiny crystals of certain chemicals, or electrically charged particles.

In the 1940s, the American physicist Vincent J. Schaefer was carrying on a program of experimentation with water vapor in a closed container kept at very low temperatures in order to duplicate cloud behavior in the laboratory. In July 1946, during a hot spell, he found it difficult to keep the temperature low enough. He placed some solid carbon dioxide (Dry Ice) in the box, in order to force down the temperature. At once the water vapor condensed into large ice crystals and the box was filled with a miniature snowstorm. Would this work on a large scale, too?

On November 13, 1946, Schaefer flew in an airplane over a cloud bank and dropped six pounds of Dry Ice into them, starting the first man-made precipitation in history. Each particle of Dry Ice was a *seed* about which a conglomerate could grow. What Schaefer had done was called *cloud seeding*. Schaefer's coworker, Bernard Vonnegut, used tiny crystals of silver iodide. These were more effective seeds than Dry Ice and could be blown into the clouds from the ground. Beginning with the 1950s, cloud seeding has been used often in attempts to break droughts or to disrupt hurricanes before they can build up to dangerous levels.

Coacervate

SCIENTISTS STUDYING the possible beginnings of life have duplicated conditions as they were thought to have existed billions of years ago. They have shown that from the very simple compounds present in the original ocean, more complicated compounds could surely have been built up. Eventually, compounds similar to those that now exist in living tissue could have formed.

But once such complicated compounds form, how do they come together to form living systems? Somehow they must have come together to form organized cells; cells simpler than any now existing, perhaps.

It happens though that, in solution, complicated molecules do not invariably remain evenly spread out. Under certain conditions, such a solution divides into two parts. In one part, the complicated molecules are concentrated, while in the other, few or none are to be found. (This is because the complicated molecules often have some sort of mutual attraction due to small electric charges here and there on the molecular structure.)

Such a separation into a part rich in large molecules and a part poor in them is called *coacervation* from a Latin word meaning "to heap up." The portion rich in the large molecules (where they are heaped up, so to speak) is the coacervate.

The Russian biochemist Alexander I. Oparin as long ago as 1935 suggested that the first cells might have formed as small droplets of coacervates. In 1958, the American biochemist Sidney W. Fox formed proteinlike compounds he called *proteinoids* by processes not involving life. He dissolved them in hot water and let the solution cool. The proteinoids formed tiny spheres of coacervates about the size of small bacteria. Fox called these *microspheres*.

These microspheres turned out to have some properties reminiscent of cells, which made the coacervate theory more attractive.

Codon

IN THE 1940s, it was discovered that nucleic acids served as models for the synthesis of particular enzyme molecules in the cell. Nucleic acid molecules are made up of long strings of units called nucleotides, while enzyme molecules are made up of long strings of units called amino acids.

If each nucleotide in the nucleic acid chain corresponded to some particular amino acid, you could imagine a neat transference of structure from nucleic acid to enzyme. The trouble was, though, that there were only four different nucleotides in nucleic acid molecules, but there were twenty different amino acids in enzyme molecules.

This, however, is not as puzzling as it sounds. There are only nine digits, but used in combinations, they can represent an infinite number of integers. Suppose a different combination of neighboring nucleotides stood for each different amino acid.

If four different nucleotides occur in any combination along a chain, there are 4×4 or 16 possible different neighboring pairs of nucleotides, and $4 \times 4 \times 4$ or 64 possible different neighboring trios. It would have to be combinations of three nucleotides that corresponded to the twenty different amino acids.

During the 1960s, biochemists determined which amino acid each of the sixty-four possible trios stood for, and this was called the *genetic code*. (Two or sometimes three or four different trios might all stand for the same amino acid.)

What does one call a particular combination of three nucleotides? The influence of the phrase *genetic code* made itself felt and the three-nucleotide unit, corresponding to a particular amino acid, was called a *codon*. The "-on" ending was borrowed from the subatomic particles which were fundamental parts of an atom, as the codon was a fundamental part of the nucleic acid molecule.

Coelacanth

ABOUT 450 MILLION YEARS AGO, fish evolved and became the dominant life form of the oceans. One large division of fish had four stubby, fleshy appendages, fringed with fins. This group of fish were therefore named *crossopterygii* from Greek words meaning "fringe fins." The crossopterygii were not as efficient in swimming as were those other fish which developed appendages that were less fleshy and bore longer fins.

Thanks to their more muscular appendages, however, the crossopterygii could manage to stump their way across land when they had to. It is thought that land animals are descended from them, but they themselves were rapidly dwindling in numbers 250 million years ago.

The crossopterygii were freshwater fish to begin with, but some forms colonized the salt sea. Certain of these are called *coelacanths*, from Greek words meaning "hollow spines," because this was one of their features. No coelacanth fossils were found to be less than 70 million years old and it was thought they had become extinct.

On December 25, 1938, however, a trawler fishing off South Africa brought up an odd fish about five feet long. A South African zoologist, J. L. B. Smith, who had a chance to examine it, recognized it at once as a living coelacanth. World War II halted the hunt for more coelacanths, but in 1952 another, of a different variety, was fished up off Madagascar. Before long, numbers of this kind of fish were found. Since they are adapted to fairly deep waters, they die soon after being brought to the surface.

The coelacanths have a double interest. First, they are living examples of a type of creature that was thought to have grown extinct with the dinosaurs. Second, they are the only known direct descendants of a type of fish that seem to have been the ancestors of land vertebrates and, therefore, of ourselves.

Coenzyme

In 1904, two English biochemists, Sir Arthur Harden and W. J. Young, studying a yeast enzyme, zymase, disrupted it into two parts. Separated, the two parts of the enzyme could not carry through the normal task of bringing about the fermentation of sugar. When the two were combined, however, enzyme activity was restored.

One of the fractions was a protein very much like the original enzyme. It was the main portion of the molecule. The other was not protein, but was, instead, a relatively small molecule that was stable to heat. It was a substance that worked with the main portion of the enzyme to help it carry out its function and so Harden and Young called it *cozymase*, where the prefix is a Latin one meaning "together with."

In general, a small molecule, necessary to enzyme function but radically different in chemical structure from the rest of the enzyme, is called a *coenzyme*.

It was soon found that many coenzymes contained atom groupings that were unusual and were not found in other tissue components. Since enzymes need be present in only small quantities to do their work, so need be coenzyme and these atom groupings. It happens that the tissues of complex animals (and those of some simple organisms, too) do not maintain a complex system of chemical machinery in order to make these atom groupings that are needed in such small quantity. Instead it seems reasonable to count on finding sufficient amounts, ready-made, in food.

It is these relatively rare but necessary atom groupings which the body must find in food that are included among vitamins.

Some enzymes have associated with them atom groupings containing metal atoms — such as cobalt, copper, zinc, or molybdenum — not often found elsewhere in tissues. They are present in organisms only in traces, and yet these *trace minerals* are essential to life.

Competitive Inhibition

EMZYMES bring about, or catalyze, a chemical reaction, making it proceed with far greater speed than would be the case in the absence of the enzyme. Each different reaction requires an enzyme of its own that works to bring about that reaction and no other. Each enzyme need be present in only tiny quantities to do its necessary work, yet if that small quantity is prevented from operating, some chemical reaction is stopped, usually with serious, sometimes fatal results.

A substance which will block an enzyme's action, or *inhibit* it (from Latin words meaning "to hold in"), will seriously disturb or kill an organism. If one molecule of the inhibiting substance will stop one molecule of the enzyme, then no more of the substance is needed than of the enzyme. A small quantity of the substance can kill and such a substance is a *poison* (from a Latin word meaning "to drink," since poisons are usually thought of as being dissolved in liquid).

One way of keeping an enzyme from working is to present it with a substance possessing a chemical structure very like some portion of itself or some chemical it normally works with, but one that is not identical. This merely similar compound competes with the natural compound for combination with the enzyme. Once it is part of the enzyme, however, it is sufficiently different from the natural compound to keep the enzyme from working. This is *competitive inhibition.*

Some poisons that work by competitive inhibition are life savers for us, since not all cells are identical. Some cells are more dependent on a particular reaction than others are. Some cells have enzymes more sensitive to a particular competition, or membranes more permeable to competing molecules. It is possible to find compounds that competitively inhibit the enzymes of bacteria without seriously affecting animal cells. Antbiotics probably work in this way and are, therefore, selective poisons.

 # Continental Drift

A GERMAN GEOLOGIST, Alfred L. Wegener, was intrigued by the fact that the eastern coast of South America looked as though it would fit the western coast of Africa. Could they once have been a single land mass that split and drifted apart? Wegener had explored Greenland and, after collecting various determinations of longitude made there, decided that Greenland had moved a mile away from Europe over the previous century.

In 1912, he advanced the theory that at one time, the major land masses of Earth had formed a single body, *Pangaea* (from Greek words meaning "all earth"), surrounded by a single ocean, *Panthalassa* (all ocean). Pangaea had broken into fragments, the modern continents, and these had very slowly drifted apart, like big chunks of granite floating on the hot, thick fluid of Earth's deeper layers. This theory is called *continental drift*.

The theory was not taken seriously to begin with. The increased interest in the sea bottoms after World War II, however, and the discovery of mountain ranges in the mid Atlantic, marked by a central rift, seemed to indicate that the Atlantic Ocean might indeed be widening.

In 1968, there came a most dramatic piece of evidence in favor of continental drift. A small piece of fossil bone, located in Antarctica, was clearly part of an amphibian animal that could not have lived in an Antarctica as cold as it is today. What's more, even if Antarctica had been warmer, the amphibian could not have crossed the stretches of salt water that separate the continent from other land masses today. Instead, one had to assume that about 120 million years ago, Antarctica had been joined to South America and Africa in a more temperate latitude; that it broke off and drifted away, carrying an animal load with it. As it drifted into the polar regions, a load of ice gradually accumulated on it and its life forms died, perhaps to remain as fossils under the ice.

Continuous Creation

ALBERT EINSTEIN worked out the first theory, in 1916, that took into account the universe as a whole. (This originated the science of *cosmology*.) Einstein began with the assumption that the universe was the same in average structure everywhere. Everywhere there was an even spreading of galaxies. This was called the *cosmological principle*.

In the 1930s, some astronomers began to think the universe had begun as a small sphere of extremely dense matter which had exploded (the *big bang* theory). The explosion set off a series of changes that finally brought the universe to its present state. This meant that the average appearance of the universe changed drastically with time.

To three astronomers in England, Thomas Gold, Fred Hoyle, and Hermann Bondi, this seemed doubtful. They wondered if it might not be that the universe, taken as a whole, was the same in average structure not only everywhere in space, but at all times throughout eternity. This they called the *perfect cosmological principle*.

But how could this be so when it was known that the galaxies were constantly receding from each other, so that the universe was gradually growing larger and larger and emptier and emptier? The three astronomers suggested, in 1948, that matter was continually being created throughout the universe, so that by the time two galaxies had doubled the distance from each other, enough matter had accumulated between them to form a new galaxy.

With old galaxies moving apart and new ones forming between, the average appearance of the universe would remain the same throughout time. This is called the *continuous creation* theory and is one that results in a *steady-state universe*.

Recent astronomical observations, however, seem to place the weight of the evidence on the side of the big bang theory.

Corona

ASSOCIATED WITH A TOTAL SOLAR ECLIPSE is a ring of pearly light that surrounds the dark circle of the eclipsed sun and stretches out as far as two or three times the sun's diameter.

Oddly enough, this is not mentioned by early observers of eclipses. Apparently the eclipse was so frightening a phenomenon, and those observing it were in so strong a panic lest the sun disappear forever, that no one really stopped to notice details. The Roman writer Plutarch mentioned something about a ring of light around the eclipsed sun in the first century A.D. The German astronomer Johannes Kepler said that such a ring had been seen about the sun in an eclipse of 1567. Finally, in 1715, the English astronomer Edmund Halley published a careful description of a solar eclipse that included the ring of light.

Even in the nineteenth century, there was some question as to whether the ring of light was part of the sun, or part of the disk of the moon, which was eclipsing the sun. By 1860, when photography was first used in connection with an eclipse, it was definitely decided that the light was the glowing of the thin outer atmosphere of the sun. Ordinarily, this glow was blotted out by the intense light of the deeper layers of the sun, but when the main body of the luminary was concealed by the moon's disk, the outer atmosphere sprang into beautiful, pearly light.

Because the outer atmosphere surrounded the dark disk of the eclipsed sun like a crown resting on the head of a king (as seen from above), it was called the *corona*, the Latin word for "crown."

The corona could only be studied during the few minutes of a total eclipse at first, but then, in 1930, the French astronomer Bernard F. Lyot designed a telescope which blocked the light of the sun and made the corona visible even when an eclipse was not taking place. He called this instrument the *coronagraph*.

Cosmogony

MEN HAVE ALWAYS BEEN INTERESTED in the question of how things began. Until modern times, they have always supposed that supernatural forces were involved; that some god or gods had created the universe.

By the eighteenth century, it was clear the universe was larger than had been thought, and by then scientists were attempting to account for the beginning of things without reliance on the supernatural. The French scholar Georges L. L. de Buffon suggested, about 1750, that the planets, including Earth, had originated from matter knocked out of the sun when it collided with some other huge heavenly body about 75,000 years before.

But then, how did the sun originate? In 1798, the French astronomer Pierre S. de Laplace supposed the solar system, both sun and planets, had originated out of a vast, whirling cloud of dust and gas.

As the astronomical horizon receded, astronomers began to wonder how stars originated, clusters of stars, the Galaxy, and eventually how the entire vast universe began.

In the 1920s, it became clear that all the galaxies were rapidly receding from each other, and this gave rise to the thought that mankind was witnessing the aftereffects of a huge explosion. In 1927, the Belgian astronomer Georges E. Lemaître suggested that to begin with, all the matter in the universe had been compressed tightly into one dense mass, which had exploded. It was the aftermath of the explosion that created the universe as we know it.

The term *cosmogony* is from Greek words meaning "birth of the universe." It is sometimes applied to the study of the origin of part of the universe — a galaxy, a star cluster, even a planetary system. Lemaître's suggestion, however, was the origin of true cosmogony, the study of the birth of the entire universe.

Cosmology

THE ANCIENT GREEKS thought the universe began as a heap of matter in utter disorder, which they called *chaos*. The creation of the universe, they felt, was the creation of order (cosmos) out of this chaos. For this reason, we use *cosmos* to mean the orderly universe we observe about us.

Over the centuries, men learned more about the universe and gained a more accurate notion of its enormous extent. In Greek times, the distance to the moon was worked out; later, in the eighteenth century, the distance of the planets; in the nineteenth century, the distance of the stars; and in the twentieth century, the distance of the outer galaxies. Astronomers know now that the universe measures billions of light-years from end to end.

How can astronomers learn to understand the workings of a universe so vast? One way is to work out the laws of science from observations on Earth and hope that they apply to all the universe. The English scientist Sir Isaac Newton worked out the *law of universal gravitation* in 1683 — *universal* because he thought it applied to the universe as a whole.

It certainly applied to the solar system, and in the nineteenth century, the German-English astronomer Sir William Herschel studied pairs of stars that circled each other and found that the law of gravitation applied to them, too. Newton's gravitation alone, however, did not quite solve all problems.

In 1916, the German-born physicist Albert Einstein advanced his *general theory of relativity*, which so interpreted gravity as to make it possible to use it to deduce the structure and behavior of the universe as a whole. For this reason, we may consider Einstein to have originated the science of *cosmology* (from Greek words meaning "study of the universe"), since he made it possible to study the universe as a whole.

Cosmonaut

IN JULES VERNE's *From the Earth to the Moon*, published shortly after the American Civil War, Americans were described as making the first flight to the moon. As time wore on, it seemed more and more likely that Americans indeed would. The United States grew to be the richest and most technically advanced nation on Earth and Americans gained a reputation for inventiveness and daring. In the early 1950s, they were making plans to hurl an artificial satellite into orbit about Earth.

The United States was in no hurry, however, for there seemed no competition. The Soviet Union announced plans of its own, but the Russians were considered technologically backward and few Americans paid attention. Yet the Russians were really interested in space flight. A Russian, Konstantin E. Tsiolkovsky, had published a thoughtful book on the subject as early as 1903.

The Soviet Union did not forget this, and since Tsiolkovsky had been born in 1857, they endeavored to have their first artificial satellite launched in 1957, his centenary. In this, they succeeded. On October 4, 1957, they launched the first *Sputnik* (a Russian word meaning "satellite") and initiated the space age.

Americans were caught by surprise and began to labor mightily to catch up to and surpass the Soviets. Eventually they did so, but not before the Soviet Union scored other firsts. On April 12, 1961, they sent Yuri A. Gagarin into orbit about Earth, so that a Soviet citizen was the first man to venture beyond the atmosphere in space flight.

A number of Americans and Soviets have now ventured out into space (with Americans the first to reach the moon). Each nation has a different name for these brave men. The Americans call them *astronauts* (from Greek words meaning "star voyagers") but the Soviets even more ambitiously call their men *cosmonauts* ("universe voyagers").

62

CPT Conservation

THE BASIC RULES of physics are a series of conservation laws, which state that some numerical property of a collection of matter cannot change as a result of any alterations taking place within that collection alone. The *law of conservation of energy*, for instance, states that in any part of the universe shielded from interaction with the rest, the total quantity of energy can neither increase nor decrease under any circumstances.

Physicists have worked out a series of conservation laws for sub-atomic particles by observing what changes never seem to take place among them. For a while, a certain property called *parity* seemed to be conserved. In 1956, however, two Chinese-American physicists, Tsung Dao Lee and Chen Ning Yang, showed that in some cases it was not conserved. This meant that there were conditions in which particles acted as though they were left-handed or right-handed.

When a conservation law proves insufficient, physicists may find ways of broadening it so that a new and more general property *is* conserved. For instance, particles occur in twin forms with opposite electric charges, or opposite magnetic fields, or both. When physicists describe this mathematically, they make use of something they called *charge conjugation*.

Charge conjugation seems to change along with parity. If an electron (with a negative electric charge) should happen to behave as though it were left-handed, its twin, the positron (with a positive charge), would behave under the same circumstances as though it were right-handed.

There is thus conservation of charge conjugation and parity taken together, or *CP conservation*. For theoretical reasons, this is linked with time. Any change in CP is reversed if time were imagined to be moving backward. The triple combination is *CPT conservation*.

Cryobiology

IT HAS BEEN KNOWN since prehistoric times that food spoiled more rapidly on warm days than on cold days. It always helped, therefore, to keep food in a cool cellar, or cave, during the warm seasons.

Man learned to pack perishables in ice, where that was available. In the twentieth century, electric refrigerators became part of the scene and the still colder electric freezers followed. This revolutionized the food habits of millions, for it made it possible to store foods a long time. In general, the lower the temperature, the longer food can be preserved.

Since World War II, the methods used for the production of very low temperatures have made it possible to refrigerate far more deeply than ever before at a reasonable cost. Liquid nitrogen, for instance, keeps temperatures below −195° C. and has no adverse effect on food. At liquid nitrogen temperatures, food, biological samples, and other perishables can be kept almost permanently. The study of preservation at such low temperatures is called *cryobiology*. The prefix "cryo-" comes from a Greek word meaning "freezing cold."

During the 1960s, indeed, it occurred to some individuals that people themselves might be so preserved. Suppose a person were dying of some incurable disease, or just of old age. At the instant before death, he might be placed in refrigeration at liquid nitrogen temperatures. Then at some time in the future, when the disease could be cured or old age reversed, he might be revived and treated.

Of course, there is as yet no method known for reviving a person frozen at liquid nitrogen temperatures, and it is questionable whether such freeze-now-revive-later procedures are psychologically and sociologically feasible. Still, societies have been founded for the purpose of doing this and they have called the techniques involved *cryonics*.

Cryogenics

UNTIL A HUNDRED YEARS AGO, there was no way of getting temperatures much lower than those that occurred in nature. About 1860, two English physicists, James P. Joule and William Thomson (later known as Lord Kelvin), expanded gas under conditions where no heat leaked into it from outside and found that its temperature dropped. This is called the *Joule-Thomson effect.*

This effect is made use of in refrigerators and air-conditioning devices. There a gas, liquefied under pressure, is allowed to evaporate and cool down everything about it. The process is repeated over and over, the compression taking place outside the system and the evaporation inside, so that heat is pumped out of the system continually. Scientists used the effect to reach lower and lower temperatures.

Ice melts at 0° C. or 273° K. (that is 273 degrees above absolute zero). In 1877, the Swiss physicist Raoul Pictet reached a temperature of 133° K., lower than any that occurs naturally anywhere on Earth. At that temperature, he produced liquid oxygen.

In 1900, the Scottish chemist James Dewar managed to reach a temperature of only 33° K. and produced liquid hydrogen by placing that gas under pressure. Finally, in 1911, the Dutch physicist Heike Kamerlingh Onnes produced 4.2° K. and liquefied the last remaining gas — helium.

By allowing liquid helium to evaporate, a temperature of 1° K. can be reached, and still lower temperatures, to within a millionth of a degree of absolute zero, can be reached by methods more subtle than the Joule-Thomson effect. (Absolute zero itself cannot be attained.)

The study of the properties of matter at liquid helium temperatures has yielded surprising and useful results. This study is called *cryogenics* from Greek words meaning "to produce freezing cold." Liquid helium temperatures are therefore called *cryogenic temperatures.*

Cryotron

THE TREND in electronics is in the direction of *miniaturization;* that is, in making a device ever smaller. The smaller a device, the less material is required to construct it and the more portable it is. One great step in this direction was the replacement of the relatively bulky radio tube by the much smaller transistor.

The transistor does not represent the most compact possible device for the delicate control of electron flow, a control that makes electronic devices possible. Two small wires would be sufficient.

This results from the fact that some metals lose all electrical resistance at very low temperatures, becoming *superconductive.* This superconductivity can be wiped out, even at very low temperatures, if a magnetic field of sufficient intensity is applied.

Consider a tiny straight wire of tantalum. At temperatures below 4.2° K., tantalum is superconducting and a current sent through that wire could continue to exist indefinitely. But suppose a spiral of niobium wire is wrapped around the first. Niobium is superconductive up to 9.2° K. and can withstand, at any temperature, a larger magnetic field than tantalum can before losing superconductivity. If current is sent through the niobium, a magnetic field is set up and can be made strong enough to wipe out the superconductivity in the tantalum, but not in the niobium itself.

In this way, the electron flow in the first wire can be delicately controlled by changing its resistance by means of an electric current through the second wire — but of course only while temperatures are maintained at those of liquid helium, very near absolute zero, the only condition under which superconductivity can exist. Such temperatures are *cryogenic* (to produce freezing cold). The two wires represent *cryogenic electronics* and are therefore termed *cryotrons,* an abbreviated form of that phrase.

Cryptozoic Eon

GEOLOGICAL HISTORY is usually divided into long periods of time on the basis of the kind of fossils that are characteristic of those periods. The oldest rocks in which fossils are to be found are of the *Cambrian era* (named for Cambria, the Latin name for the region now known as Wales — where such rocks were first studied). The Cambrian rocks are up to 600 million years old, and anything older, with no fossils in it, was at one time simply considered part of the *pre-Cambrian era.*

In recent years, however, it has become more and more evident that there are clear traces of life in pre-Cambrian rocks. There aren't obvious fossils, to be sure, but there are *microfossils* and organic chemicals that seem to have been produced by one-celled creatures. For that reason, all the geological eras from the Cambrian to the present day are now grouped as a single *eon,* the *Phanerozoic eon* (from Greek words meaning "visible life," thus implying the existence of previous life not so easily visible).

The pre-Cambrian era has now become the *Cryptozoic eon* (hidden life), divided into two sections: an earlier *Archeozoic era* (ancient life) and a later *Proterozoic era* (early life).

The division between the Cryptozoic eon and the Phanerozoic eon is extraordinarily sharp. At one moment in time, so to speak, there are no fossils at all above the microscopic level, and at the next there are elaborate organisms of a dozen different basic types. Such a sharp division in the geologic record is called an *unconformity.*

Unconformities usually imply some sharp change in conditions to which Earth is exposed. Explanations for the unconformity between the two eons have ranged from the rapid production of oxygen after the development of photosynthesis, which made elaborate life possible, to Earth's capture of the moon, which created huge tides that wiped out the earlier record.

Cybernetics

THROUGH THE PRINCIPLE of feedback, any process is guided by the difference between the actual state of affairs at the moment and the desired state of affairs. The rate of approaching the desired state decreases as the difference decreases. When the desired state is reached, so that there is zero difference between actual state and desired state, the change stops.

In 1868, a French engineer, Léon Farcot, used this principle to invent an automatic control for a steam-operated ship's rudder. As the rudder approached the desired position, it automatically tightened a steam valve which made the rudder move more slowly. By the time the desired position was reached, the steam pressure was shut off and the rudder moved no more. If the rudder moved out of place, the valve opened and steam pushed it back. Farcot called his device a *servomechanism* (slave machine) because it was as though a slave sat there constantly adjusting the position. Since then, more and more machinery has been designed to adjust itself automatically. In 1946, an American engineer, D. S. Harder, coined the word *automation* to describe such adjustment.

Electronic devices made automation more delicate, and radio beams extended the effect over long distances. The German buzz bomb of World War II was essentially a flying servomechanism, and this quickly escalated to the horror of intercontinental missiles with nuclear warheads. Space exploration would be impossible without automated devices.

In the 1940s, the American mathematician Norbert Wiener worked out some of the fundamental mathematical relationships involved in the handling of feedback. He named this branch of study *cybernetics*, from the Greek word for "helmsman." A helmsman, after all, controlled a rudder by constantly observing its position, and this is what automated cybernetic devices did too, but more tirelessly and precisely.

Cyclic-AMP

In 1885, the Swiss chemist Albrecht Kossel isolated a compound which eventually turned out to be a component of some of the most important substances in living tissue. He isolated it from pancreas, the second largest gland in the body, and so he named it *adenine* from the Greek word for "gland."

In nucleic acids, adenine is found combined with a sugar called ribose and the combination was named by using letters from both names: *adenosine*. In addition, a phosphate group (containing atoms of the element phosphorus) is usually attached. Sometimes two, and even three, are attached. The resulting compounds are *adenosine monophosphate*, *adenosine diphosphate*, and *adenosine triphosphate*, the prefixes "mono-," "di-," and "tri-" coming from Greek words for "one," "two," and "three," respectively. The names are often abbreviated as *AMP*, *ADP*, and *ATP*, respectively.

ATP, in particular, is of key importance. It is involved in almost every kind of reaction which yields the energy that can be used to manufacture large molecules out of small ones.

In 1960, the American biochemist Earl W. Sutherland, Jr., discovered a variety of AMP in which the single phosphate group was attached to the adenosine portion of the molecule in two different places. Thanks to this, the molecules possessed atoms linked together to form a closed ring, or cycle. For this reason, Sutherland called it *cyclic-AMP*.

Cyclic-AMP was found to be widely spread in tissue and to have a pronounced effect on the activity of many different enzymes and cell processes. It casts new light on how hormones achieve their results — something which has been hitherto mysterious. It may be that different hormones affect the production or destruction of cyclic-AMP in different ways and that this in turn affects the cell chemistry in some crucial fashion.

69

Cytochrome

BEGINNING IN 1885, certain cell components of unknown function were detected through their ability to absorb light of certain wavelengths. This meant that, if isolated, those cell components would be colored. In 1925, the Russian-British biochemist David Keilin named these substances *cytochromes*, from Greek words meaning "cell colors."

Keilin distinguished three different cytochromes by differences in the fashion in which they absorbed light and labeled them *a*, *b*, and *c*. Since then, each of these three has been found to consist of two or more varieties.

All the cytochromes are proteins possessing an iron-containing portion like the *heme* in the well-known protein hemoglobin. Like hemoglobin, the cytochromes can attach oxygen molecules to themselves.

There is a difference, though. The hemoglobin molecules serve merely as a chemical-transport system, carrying oxygen molecules from lungs to tissues. To do so, they attach oxygen to themselves loosely, without changing the chemical nature of the iron atoms to which they are attached. In the case of the cytochromes, the oxygen molecules are attached more tightly and the iron atoms change character, each losing an electron on combining with the oxygen (and regaining the electron when giving up the oxygen).

In the 1940s, it came to be understood that the various cytochromes formed a chain. Oxygen atoms passed from one to the other, energy being liberated at each step in small quantities that could be usefully stored by the body, until finally each oxygen atom was combined with a pair of hydrogen atoms obtained from fragments of food molecules that had been absorbed into the body.

Every cell that makes use of oxygen contains cytochromes and these form part of the structure of small cell components called mitochondria.

DDT

THERE ARE nearly a million different species of insects known and this is far more than the total number of different species of all other animals. Of all these species, only about 3000 are nuisances to man but these include mosquitoes, flies, fleas, lice, wasps, hornets, weevils, cockroaches, carpet beetles, and the various insect species that live on the plant life or harass the animal life or damage the manufactured objects that man is trying to preserve for his own use.

As a result, men have long tried to kill insects in every way possible. As late as the 1930s, copper- and arsenic-containing poisons were used in sprays, but these gradually poisoned the soil. In 1935, a Swiss chemist, Paul Mueller, began a search for some organic chemical that would kill insects but not other forms of life and that would be cheap, stable, and odorless.

Certain chlorine-containing compounds showed promise. In September 1939, Mueller tried one such compound called *dichlorodiphenyltrichloroethane*, a compound which had been known since 1873 — and it worked!

The name was soon simplified to the letters that started its first, fourth, and sixth syllables and it became *DDT*. In 1942, it was produced commercially in the United States, but was reserved for military use. In Janury 1944, it was used to kill body lice in Naples and prevent a typhus epidemic. It was similarly used in Japan in late 1945.

After the war, DDT came to be used against insects everywhere. It was by far the best-known and most popular *insecticide* (from Latin words meaning "insect killer"). As the years went by, it was recognized as harmful in the long run to creatures other than insects and as *too* stable, for it lingered in the soil and in living tissues. By the 1970s, government after government began to control its use.

Dendrochronology

IN STUDYING THE PAST, men are interested in when events took place; whether one event took place after another, or before, or at the same time. For recent periods, there are records to consult. For times before the invention of writing (which is only about 5000 years old at most, and not nearly as old in most parts of the world), something else must be found.

Can natural events do the writing? In some ways, nature is so regular as to be useless. The sun rises and sets in a fixed pattern; the moon changes its phases; the seasons come and go. The rain, however, varies, with plenty of rain in some years and drought in others.

The American astronomer Andrew E. Douglass, working in the desert regions of the American southwest in the early decades of the twentieth century, considered tree growth. Each growing season produces a layer of light wood about the trunk. The thin, dark layers between are the *tree rings*. A rainy season means considerable growth and a wide space between two rings; a dry season means little growth and a narrow space. A twenty-year-old tree would demonstrate a twenty-year pattern of rainfall that would not be exactly duplicated in any other twenty-year period.

Douglass studied many trees and found the pattern overlapping; the pattern in the early years of one tree would be like that in the later years of an older tree. He used wood salvaged from ancient buildings and carried a continuous pattern back over a thousand years. A piece of wood from some man-made structure could be dated by fitting the pattern of its rings somewhere along Douglass' continuous pattern. Thus, one could tell when a prehistoric structure was built.

This device for dating early events was named *dendrochronology* by Douglass, from Greek words meaning "telling time by trees."

Digital Computer

THE FINGERS were the first device used by man for counting, and for simple computations in adding and subtracting. *Digit* (from the Latin word for "finger") therefore means both a finger and any number below ten. The fingers are thus the first *digital computer*.

Any mechanical device that represents whole numbers the way that fingers do, and can be manipulated as numbers can, so as to give correct answers to numerical problems, is also a digital computer. The abacus, in which numbers are represented by pebbles in grooves, or disks on wires, is a convenient digital computer.

In 1642, the French mathematician Blaise Pascal invented a mechanical device of wheels and gears. Each wheel had ten positions, one for each digit from 0 and 9. When the first wheel reached 9 and passed to 0, it engaged the second wheel, which turned to 1. When the second wheel turned as far as 0, the third wheel was engaged, and so on. In this way, numbers could be added and subtracted mechanically. In 1674, the German mathematician Gottfried von Leibnitz arranged wheels and gears so that multiplication and division were also made automatic.

In 1850, an American inventor, D. D. Parmalee, pushed marked keys to turn the wheels and this was the cash register.

Such mechanical devices were improved by electronic techniques. An electronic computer was first built during World War II, according to the plans of the American engineer Vannevar Bush, in which electric currents replaced mechanical gears. Such electronic digital computers were rapidly and vastly improved. Transistors took the place of vacuum tubes and reduced computer size; sophisticated "memories" were added; "languages" were developed for direction. Now computers can far outstrip the speed of the human mind in any purely mechanical computation for which a computer can be given the necessary complete and detailed instructions.

Dipole Moment

MANY SUBATOMIC PARTICLES carry an electric charge, either positive or negative. Atoms are made up of two kinds of such particles: protons with a positive charge, and electrons with negative charge. The number of each kind in a complete atom is equal. What's more, both kinds of particles are distributed evenly about the center of the atom. The average position of the positive charges and of the negative charges are both exactly at the center of the atom. The effects of each cancel and the complete atom behaves as though it has no charge.

When two similar atoms join, as two chlorine atoms do in forming the chlorine molecule, they share electrons equally. The two types of charge are still distributed evenly about the center of the molecule. When two unlike atoms cling together, however, and share electrons, one usually has a stronger hold on the electrons than the other does. The average position of the negative charge shifts away from the center of the molecule and toward the atom with the stronger hold.

In the latter case, the average position of the negative charge is in a different place from that of the positive charge. There is a separation of the two charges. We can say there is a *positive pole* in one place and a *negative pole* in another. (The terms are an analogy to the north and south poles in magnets.) Because there are two poles, such a molecule is a *dipole*.

In an electric field, such a dipole moves so as to have the line connecting the poles parallel to the direction of the field. The readiness with which it moves depends on the size of the charges and the amount by which they are separated. This readiness is the *dipole moment*, where *moment* is a form of the word *movement*.

This notion, first worked out in 1912 by the Dutch chemist Peter J. W. Debye, has helped explain the behavior of substances such as water.

Domains, Magnetic

ATOMS ARE MADE UP of electrically charged particles, and any electric charge always has an associated magnetic field. All matter, therefore, has the potentiality of displaying magnetic effects, and yet most types of matter scarcely do. Iron, steel, and related materials are much more easily affected by magnetic forces than other substances commonly found in nature. The particular type of magnetism displayed by iron is therefore called *ferromagnetism*, the prefix coming from the Latin word for "iron."

An explanation of iron's peculiar relationship to magnetism was advanced in 1907 by the French physicist Pierre E. Weiss. He suggested that, in general, the tiny atomic magnets in matter were oriented every which way so that the effects were neutralized. In iron, on the other hand, there were microscopic regions over which many billions of atoms orient themselves in such a way that all the tiny magnetic fields are in the same direction. Each tiny region is therefore a much stronger magnet than a similar region of any other substance would be. Such a region of strong magnetic field is called a *magnetic domain*.

In ordinary iron, the magnetic domains are oriented every which way so that the iron is not particularly magnetic. The domains are much easier to orient than individual atoms are, however. Earth's magnetic field produces such an orientation in a ferric oxide called lodestone, and this can produce orientation in metallic iron or steel.

If a ferromagnetic substance is ground into particles smaller than the individual domains making it up, each particle will consist of a single domain. If these are suspended in liquid plastic, they can easily be aligned by the influence of a magnet. If the plastic is then allowed to solidify, a particularly strong magnet results; one, moreover, that can be easily machined into any desired shape.

Echolocation

In 1793, the Italian biologist Lazzaro Spallanzani grew interested in the manner in which bats could flit about in the dark, avoiding obstacles. Could they see in the dark? He blinded some bats and found that they could still fly without difficulty and without blundering into obstacles. However, when he plugged their ears so that they could not hear, they were helpless and blundered into obstacles even though their eyes were open and working. Spallanzani had no explanation for this, and could only record the observation.

Studies in the twentieth century showed that bats constantly emit very high-pitched squeaks. Some of the sounds they make are *ultrasonic* (from Latin words meaning "beyond sound"). The sounds were of such high frequency and short wavelength, and therefore so highly pitched, that the human ear could not detect them.

Sound is reflected from objects, producing an echo. The size of the object required to produce an echo depends on the wavelength of the sound. Ordinary sound has such long wavelengths that it takes large objects like walls to produce an echo. Ultrasonic sound has such short wavelengths that small objects, even twigs or insects, could produce some echo.

The bat, emitting its high squeaks, listens for echoes with its large, sensitive ears. From the direction of the echo and from the time it takes the sound to reach the obstacle and return, it can tell the direction and distance, and even the nature of the obstacle. It can avoid a twig and snatch at an insect.

It is possible that dolphins also use such a system of *echolocation*, using somewhat deeper sounds and detecting larger objects, such as fish. The guacharo, a cave-dwelling bird of Venezuela, may also use echolocation.

Ecology

GENERALLY, mankind has not concerned itself with other forms of life except as they suited its purpose. Useful plants or animals were preserved, cultivated, or herded. Other animals were killed, sometimes for sport, even when they were not dangerous. Still others were ignored.

Men have come to realize, however, that organisms (even ourselves) do not live in isolation. Each depends on others. A species may even depend for its well-being on another species that preys upon it. Deer are better off because mountain lions exist, for instance.

In some areas, the mountain lions were killed off by hunters because they preyed on domestic animals. The deer, freed of the menace, multiplied and outran their food supply. Many of them starved, and in the end, there were fewer and weaker deer than before. Since mountain lions usually catch and kill the older and sicker deer, they serve to keep the deer herds younger and stronger than would be the case otherwise.

Man's interference often upsets the *balance of nature* and has created deserts. It has led to the extinction of harmless creatures preyed on by animals carelessly introduced into their areas by man.

In fact, as man's numbers grow and his technological ability to change the environment increases, he is more and more rapidly and extensively changing the balance of nature. With this in mind, scientists are beginning to study the interrelationships of life forms with each other and with the environment, in order to learn best how to stop and reverse man's damaging effect.

The prefix "eco-" is from a Greek word meaning "house" and it can be stretched to the environment generally, for that is the house of life, so to speak. Consequently, *ecology* is the name given to the newly important study of the interrelationships of life forms among themselves and with the environment.

77

Ecosphere

CONSIDERING THE VASTNESS of the universe, it seems possible that Earth might not be unique, and that there are other worlds with life — even intelligent life.

In the 1930s, the solar system was thought to have been formed through the close approach of two stars, with mutual gravitational pull tearing out matter from each to form the planets. If this were so, planetary systems would be rare, since stars are spread so widely apart that they would almost never approach each other in this fashion.

In the 1940s, the German astronomer Carl von Weizsäcker advanced a theory of planetary formation from an original dust cloud that made it seem that almost any star would be accompanied by planets. And since then, some nearby stars have been shown to have planets.

Furthermore, investigation into the origins of life on Earth have shown that, given Earth-like conditions, the formation of life is almost inevitable. Consequently, any planet with the proper mass, temperature, and chemistry ought to develop life.

What, then, are the requirements for the existence of Earth-like planets? A particular star ought to have some region in its neighborhood where a planet would get just enough radiation to maintain the conditions necessary for life — temperatures at which water would be a liquid, for instance. This life-suitable region, distributed in a spherical shell all about the star, is the *ecosphere* (where "eco-" is from the Greek word for "house" and therefore expresses a place man can inhabit).

The American astronomer Stephen Dole, considering the ecospheres of different types of stars and other information as well, has estimated there may be as many as 640,000,000 planets in our galaxy alone, which might be suitable for Earth-like life.

Elastomer

COMPLICATED MOLECULES can, in some cases, be easily broken down to other molecules much simpler in structure. It was as though the more complex molecule is built up of chains of one or more much simpler molecules. In 1830, the Swedish chemist Jöns J. Berzelius suggested that such a chain molecule be called a *polymer* from Greek words meaning "many parts." The single parts of which it was a chain would then be called *monomers* (one part).

Thus, large starch molecules can be broken up and shown to consist of chains of small glucose molecules. Protein molecules consist of chains of twenty different, but related, amino acids. Rubber molecules consist of chains of a hydrocarbon called *isoprene* (a made-up name of no meaning).

Chemists found that simple compounds often hooked together in chains, even with very little encouragement. The simple compounds *polymerized*. Generally, this resulted in a gummy, useless substance that chemists tried to avoid.

In the twentieth century, however, ways were discovered for deliberately creating polymers that might be useful because they were strong, stable, and could be molded. These came to be called *plastics* because something that is plastic can be molded.

With the coming of the automobile and its rubber tires, there was a demand for artificial rubberlike polymers that might be cheaper or more available than natural rubber. By the 1940s, *synthetic rubber* was in production.

Rubber has the ability to deform its shape and then spring back to its original shape. This is called *elasticity* (from a Greek word meaning "springiness"). Synthetic rubbers were, therefore, *elastic polymers*, and this phrase was shortened to *elastomers*.

79

Electrocardiogram

THE REGULAR RHYTHM of the heartbeat is maintained by a periodic electrical change that travels along the heart muscle. If it were possible to follow this change, one might detect abnormalities in the heart action impossible to catch by merely listening to the heartbeat.

It is impractical to place electrodes directly on the heart for some routine examination, but the tissues conduct electricity, and electrodes on the skin would do the trick if a device delicate enough to detect very tiny changes could be developed.

Such a device was first perfected by a Dutch physiologist, Willem Einthoven, in 1903. He made use of a very fine fiber of quartz that was silvered to allow it to conduct a current. Even tiny electrical changes caused noticeable deflections of the fiber. These movements could guide a pen which would mark out an irregular line on a slowly unrolling length of graph paper. The result is an *electrocardiogram,* from Greek words meaning "written record of heart electricity."

The heart is not the only organ that works by rhythmic changes in electrical conditions. The organs most intimately connected with electric pulses are the nerves, and it would therefore not be the least bit surprising to find that the brain was a source of varying electrical changes. In 1875, an English physiologist, Richard Caton, applied electrodes directly to the living brain of a dog and could just barely detect tiny currents.

Such an experiment on humans seems unfeasible, but in 1924, an Austrian psychiatrist, Hans Berger, placed electrodes against the human scalp and, by using a very delicate detecting device, found he could just record electrical changes. He published his results in 1929, and improvements since then have made it easy to obtain and study *electroencephalograms* (written records of brain electricity).

Electroluminescence

LIGHT IS ALWAYS PRODUCED by hot objects. The sun, an electric-bulb filament, burning gases from wood, coal, oil, or a candle, all give off light because they are at temperatures of over 600° C. Such high-temperature light is called *incandescence* (from Latin words meaning "to become hot enough to glow").

Nevertheless, light can also be formed in the absence of high temperature. Chemical reactions within the tissues of a firefly, for instance, produce small quantities of light at ordinary temperatures. This is sometimes called *cold light*, but the more formal name is *luminescence* (from Latin words meaning "to become light").

Some substances can give off energy in the form of light after they have absorbed energy in some other form. For instance, the compound zinc sulfide, when properly prepared, will absorb energy from an electric field and then give it off again in the form of light. It will glow while remaining at room temperature, and this is called *electroluminescence*. This was first studied by the French physicist Georges Destriau in 1936.

In ordinary light bulbs, which work by incandescence, very little of the energy used is radiated in the form of visible light. Most is in the form of invisible infrared radiation which can be felt as heat. An electroluminescent powder, prepared over a sheet of glass, can be made to glow softly, producing only visible light and very little infrared. Much less electrical energy is consumed to produce a given amount of light. The electroluminescent panel is much more expensive to prepare than a light bulb is, however, and it will deteriorate rather easily under improper conditions of temperature and humidity.

Substances, like zinc sulfide, that will exhibit electroluminescence are sometimes called *electroluminors*.

Electrophoresis

GIANT PROTEIN MOLECULES come in many varieties, and protein extracts from tissues usually contain a large number of very similar molecules which, nevertheless, can behave quite differently within the body. Attempts to separate and isolate the different components of these complex mixtures failed when ordinary chemical methods were used.

However, each protein molecule has negative and positive electric charges distributed across itself and, as a result, will be attracted in one direction or another by an electric field. The direction and intensity of the attraction will depend on the pattern of charges on the molecule and this is different for each molecule, even when two of them are otherwise very similar. This was first pointed out, in 1899, by the English biologist Sir William B. Hardy.

If an electric current is passed through a solution of proteins, some molecules will travel toward one electrode, some toward the other, each at its own characteristic speed. As a result, the protein molecules separate and the number and identity of the components making up the mixture can be worked out. This procedure is *electrophoresis*, from Greek words meaning "borne by electricity."

This way of separating proteins did not become practical until methods were worked out for detecting fine degrees of difference from point to point in the solution, as the nature of the mixture changed with the gradual separation of different molecules. In 1937, the Swedish chemist Arne W. K. Tiselius devised a set of tubes that could be put together at specially ground joints, into a rectangular U. By the use of certain lenses he could follow differences in the protein mixture by changes in their ability to bend light rays, and by taking the U tube apart, he could trap one component or another in the different sections.

Electrophoresis has been used to detect tiny changes in blood chemistry in the course of various diseases.

Entropy

ENERGY CAN BE CONVERTED into work, and the law of conservation of energy states that the quantity of energy in the universe must stay forever the same. Can one then convert energy into work endlessly? Since energy is never destroyed, can it be converted into work over and over?

In 1824, a French physicist, Nicolas L. S. Carnot, showed that in order to produce work, heat energy had to be unevenly distributed through a system. There had to be a greater than average concentration in one part and a smaller than average concentration in another. The amount of work that could be obtained depended on the difference in concentration. While work was produced, the difference in concentration evened out. When the energy was spread uniformly, no more work could be obtained, even though all the energy was still there.

In 1850, a German physicist, Rudolf J. E. Clausius, made this general and applied it to all forms of energy — not just to heat. In the universe as a whole, he pointed out, there are differences in energy concentration. Gradually, over the eons, the differences are evening out, so that the amount of work it will be possible to obtain will grow less and less forever, until all the energy is evened out and no more work is possible. This is the *second law of thermodynamics*, the conservation of energy being the *first law of thermodynamics*.

Clausius worked out a particular relationship of heat and temperature which, he showed, always increased in value as the differences in energy concentration evened out. He called this relationship *entropy* for some reason. (It comes from Greek words meaning "to turn out," which seem unconnected with the case.) The second law of thermodynamics states that the entropy of the universe is always increasing.

With the discovery of quasars and other mysterious energy sources in the universe, though, astronomers now wonder if the second law really holds everywhere, and under all conditions.

Escape Velocity

When an object is thrown into the air, the pull of gravity slows it more and more until it comes to a momentary halt, then begins to drop toward Earth again. If it is thrown with greater force it moves upward more speedily and it takes longer for the pull of gravity to slow it to a halt. The object moves higher before it begins to drop. The more forcefully it is thrown upward, the higher its topmost point.

As it happens, the pull of gravity weakens as an object moves farther and farther from Earth's center. An object hurtled upward with such force that it attains a height of many miles finds the pull of gravity in the upper reaches of its flight to be significantly weaker. It is less effectively slowed up there and reaches a greater height than would otherwise be expected.

Suppose a body were hurtled upward with such force that by the time it had lost half its upward velocity, it was in a region of space where Earth's gravitational pull was only half what it was at the surface. By the time its upward velocity was a quarter of the original, the gravitational pull would also be only a quarter of what it was to begin with. Under these conditions, the object would move upward more and more slowly, but weakening gravitational pull would never bring it to a complete halt. It would never return to Earth but escape permanently into space.

The minimum initial velocity which will accomplish this is the *escape velocity*. On Earth it is seven miles per second. On more massive planets, like Jupiter, the escape velocity is higher, and on less massive ones, like Mercury, it is lower. The matter of escape velocity, known since the time of Sir Isaac Newton in the 1680s, has become of particular importance since 1959, when the Soviet Union hurled the first object into space at greater than escape velocity and sent it past the moon on a trip forever away from Earth.

Eukaryote

In 1831, the Scottish botanist Robert Brown detected a noticeable oval region within the plant cells he was studying, and because it seemed to him to be located within the cell, like the kernel within a nut, he called it the *nucleus* of the cell, from a Latin word meaning "little nut."

Virtually all cells have nuclei and although they represent a small portion of the total volume of the cell, they are a vital portion without which the cell could not survive. The nucleus contains the chromosomes that include the genes that control the formation of specific proteins. The genes determine the nature of the cell machinery, and govern the inheritance of that machinery in the course of cell division and also in the course of reproduction of entire organisms.

The separation of the reproductive and hereditary machinery of the cell into a special section at the center of the cell seems to represent a move in the direction of security and efficiency. All cells that possess a nucleus are called *eukaryotes*. This is from Greek words meaning "well nucleated."

The cells of our own tissues and of the tissues of all multicellular plants and animals are eukaryotes. The larger unicellular plants and animals are also eukaryotes.

Still, there must have been a time when the primitive cells of eons past had not yet developed the efficient isolation of the nucleus; when the reproductive and hereditary machinery was as yet scattered throughout the body of the cell. As a matter of fact, remnants of that early kind of life still exist. Bacteria, for instance, do not possess distinct nuclei, but have nuclear material distributed throughout the cell. They are *prekaryotes* (before the nucleus). Blue green algae are examples of prekaryotes that possess chlorophyll.

Eutrophication

LIFE ON EARTH depends on a balance among different species. One species serves as food for others. The wastes of one species are the fertilizing substance of others.

The activities of mankind can serve to upset this balance. For instance, man's chemical fertilizers and his detergents contain nitrates and phosphates which he gets out of the minerals of the soil. In the soil, these would slowly be washed into the freshwater lakes and into the ocean. Man, when he is through with these materials, quickly dumps them into Earth's waters.

In the ocean, there has been enough water to survive the shock, but in the confined waters of a lake, the sudden influx of nitrates and phosphates supplies a huge bonanza for various bacteria. These multiply tremendously as a result of an increased supply of minerals needed for their tissues. In the process, the multiplying bacteria consume the oxygen dissolved in water at an abnormally high rate. The oxygen content of the water consequently drops and the animal life present begin to suffocate and die. They decay, which means a further growth in bacterial activity and a further reduction in oxygen supply. Even the bacteria finally suffocate and die.

The algae, one-celled plants, don't need oxygen, and they grow brilliantly. With no decay bacteria to break them down when they die, they form a green scum on the water, while undecayed sewage collects in the lake and causes the body of water to stink and silt up.

This process, whereby the addition of fertilizing substances causes a wild, initial growth of some forms of life, followed by the death of all or nearly all, is called *eutrophication* from Greek words meaning "good nourishment." Of the large lakes in the U.S., the shallow Lake Erie seems furthest eutrophicized, and there is some question whether it can now be saved.

Exchange Forces

IN THE EARLY 1930s, it was discovered that the atomic nucleus consisted of positively charged protons and uncharged neutrons. Positive electric charges repel each other with enormous force when crowded together in the tiny space of the nucleus, however. What held the nucleus in place? In 1932, the German physicist Werner Heisenberg suggested that the nucleus held together because neutrons and protons were exchanging electric charge very rapidly.

Such an exchange might act to keep the particles together even against the repulsion of the electric charges — and yet not allow them to get too close together. The situation would resemble two boys playing catch. In order to throw the ball back and forth, the boys must not be too close or the game will be no fun; nor too far apart, or they will not be able to reach each other. As long as they play catch, therefore, they must remain at a certain distance from each other, and anyone watching from a distance who could not see the ball might be puzzled at seeing the boys move back and forth and roundabout yet always maintain a certain distance.

Heisenberg called this an *exchange force*, because it seemed to be a force of attraction resulting from exchanges of charge. As it turned out, though, the theory did not allow a strong enough attraction to account for the nucleus. In 1935, the Japanese physicist Hideki Yukawa introduced the meson. When that particle, with mass, was exchanged, rather than electric charge alone, the exchange force became strong enough.

It is possible that other forces which seem to extend across empty space can be accounted for by the continual exchange of particles. Electromagnetic forces may result from the emission and absorption of photons; gravitational forces from the emission and absorption of gravitons; the weak nuclear force by the emission and absorption of W-bosons.

Exobiology

IN THE DAYS when Earth seemed to man to be the only world in existence, Earth was naturally assumed to be the only abode of life. When
it came to be known that the moon and the planets were worlds and
there were other worlds that couldn't be seen without a telescope
(like the satellites of the other planets), and that there might even be,
and very likely were, planets circling other stars, the question arose as
to whether there was life on those other worlds.

The first impulse was to believe there was. Might not there be intelligent creatures on the moon? As late as the 1830s, a series of hoax
articles in the *New York Sun*, telling about the discovery of intelligent beings on the moon, was believed by millions. In 1877, the Italian astronomer Giovanni V. Schiaparelli noted markings on the planet
Mars that came to be thought of as *canals*. Many thought these might
have been formed by intelligent creatures.

As the twentieth century advanced, however, such notions faded.
The worlds of the solar system, except for Earth itself, were found to
lack the environment necessary for life as we know it. They were too
hot or too cold or lacked water or oxygen.

Nevertheless, astronomers wondered whether some life forms
(quite different from our own) might not adapt themselves to the
conditions on other planets, even though we might find them unsuit-
able for ourselves. They also wondered if planets of other suns, with
Earthlike conditions, might not exist. The study of life elsewhere
than on Earth was given the name *exobiology* (study of life outside)
by the American biologist Joshua Lederberg.

It is, however, a science without a subject, for there is as yet no
actual evidence for the existence of life anywhere outside Earth. An
alternative name for the science is *xenobiology* (study of stranger
life), so the science with no subject has two names.

Fallout, Radioactive

WHEN THE FIRST NUCLEAR BOMBS were exploded in 1945, the attention of the public was caught by the enormous effects of the blast and heat. The bombs were compared in their destructive effect to so many thousands of tons, or *kilotons*, of dynamite. Eventually, the comparison was to millions of tons, or *megatons*.

It became apparent, though, that there were effects of the nuclear bomb which no dynamite explosion, however great, could duplicate, effects that were more dangerous than either blast or heat.

The nuclear bomb explosion is produced by the sudden breakup or *fission* of vast numbers of uranium atoms. The breakup not only liberates energy (which produces blast and heat) but also produces uranium atom fragments, most of which are intensely radioactive. These *fission products* are carried upward in the mushroom cloud and are blown by the wind for varying distances, during which time they gradually fall out of the air to come to rest on the surface of Earth. This represents *radioactive fallout*.

If a nuclear bomb is small and is exploded at ground level, the fission products are attached to soil particles and settle out quickly within a hundred miles of the blast. Large bombs exploded in the open air, however, blast fission products high into the stratosphere, where they may drift all around Earth, gradually settling out over a space of years. In this way, nuclear bombs can poison and pollute the entire atmosphere, ocean, and soil of Earth.

It is chiefly for this reason that all-out nuclear warfare may destroy everybody and produce victory for no one. It is because of the insidious poisoning by fallout, even without war, that in 1963 the United States, the Soviet Union, and Great Britain agreed to test no more nuclear bombs by explosion in the open air.

Feedback

WE ARE SO USED to certain intricate abilities of our bodies that we take them for granted.

Suppose a pencil is on the table and you reach for it. Your hand goes to it unerringly, grasps it, and brings it back. Yet the act is a complicated one. As your hand moves out to the pencil, its motion must slow down as the pencil is approached so that by the time the fingers touch the pencil, the hand is no longer moving. The fingers must begin to close before the pencil is touched so that by the time you do touch it, only the smallest possible movement of fingers is left and you can begin withdrawing at once.

To do all this you must look at your hand and the pencil and make continuous corrections. If it looks as though the hand is slowing up too much, it must be speeded; if it is going too quickly, it must be slowed; if it is veering off to one side or another, it must be brought back to the desired path. All this is done so automatically, delicately, and quickly that you are not conscious of doing it at all and seem to make but one smooth and simple motion.

But suppose you look at the pencil, memorize its position, then close your eyes and reach. The chances are you will have to grope a bit. Some people, with brain damage, are unable to make the necessary corrections properly, and make wild motions when they try to pick up the pencil, overshooting and undershooting the mark.

Under normal conditions, the position of the moving hand and the pencil is fed back by the eyes to the brain center controlling the hand motion. It is this *feedback* that makes things work so well.

The principle of feedback works in the mechanical world, too. A thermostat works properly because it can constantly sense the actual temperature of the system whose temperature it is regulating.

Fermi

An ATOM is about 10^{-10} meters in diameter. That means that 10^{10} (or ten billion) atoms placed side by side will stretch across a meter.

Within the atom is the nucleus which contains almost all the mass of an atom, but which is far tinier still. It takes about 100,000 nuclei placed side by side to stretch across a single atom. The atomic nucleus is about 10^{-15} meters in diameter. It would take 10^{15} nuclei (a quadrillion of them) placed side by side to stretch across a meter.

The unit, 10^{-15} meters, is therefore a convenient one for measuring diameters of subatomic particles.

It also comes up in another respect. Subatomic particles freed in the course of energetic nuclear reactions move at almost the speed of light (3×10^8 meters per second). The most unstable particles only last for about 10^{-23} seconds (ten trillion-trillionths of a second) before breaking down. In 10^{-23} seconds, moving at nearly the speed of light, particles move only a little more than 10^{-15} meters, so nuclear physicists must frequently use that distance in their work.

Enrico Fermi was an Italian physicist who first studied the action of uranium under neutron bombardment; studies that eventually led to the discovery of uranium fission. He emigrated to the United States in 1938 and during World War II he was one of the leaders in the development of the nuclear bomb. He died of cancer in 1954 when he was only 53. In his honor, the unit, 10^{-15} meters, was named the *fermi*, so that subatomic particles are said to be so many fermis wide and very unstable particles are said to travel so many fermis before breaking down.

In 1962, the unit, 10^{-15} meters, was officially named the *femtometer*.

Ferromagnetism

EVERY ATOM contains electrically charged particles and each gives rise to a tiny magnetic field. The charged particles in some atoms are so arranged that their magnetic fields tend to cancel out and the atom as a whole shows no magnetic effect. In other atoms, the magnetic fields do not balance and the atom as a whole acts like a tiny magnet.

If a collection of such atoms is placed with their magnetic fields lined up in the same direction, the total effect can be quite strong. In most substances, such lining up is not possible at ordinary temperatures. In a few substances, however, there are small regions called *magnetic domains* within which all the atoms are lined up. Usually, these magnetic domains are oriented every which way, but they can be lined up more easily than individual atoms can be. With the domains lined up, a strong magnet is produced.

Natural magnets, with domains lined up by Earth's magnetic field, are found and these can be used to form still stronger magnets. Strong magnetic effects are found most commonly in iron and in cobalt and nickel, metals closely related to iron. This strong magnetic effect is called *ferromagnetism*, therefore, from the Latin word for "iron."

Ferromagnetism vanishes at high temperatures where individual atoms vibrate so strongly as to lose their alignment. On the other hand, certain substances, not ordinarily ferromagnetic, can become so at low temperatures. Nickel is no longer ferromagnetic at a temperature over 356° C., while the metal dysprosium becomes ferromagnetic below –188° C. The French physicist Pierre Curie first discovered this relationship of ferromagnetism and temperature in 1895. The temperature below which ferromagnetism exists is therefore called the *Curie temperature* of a substance.

Field-Ion Microscope

In the seventeenth century, the microscope was invented and mankind was able to enter the world of the invisibly small. During the next three centuries, the microscope was improved and objects the size of tiny bacteria could be seen with ever greater clarity. However, objects that were smaller than a single wavelength of visible light could not be seen clearly no matter how perfectly an ordinary microscope was designed.

In the 1930s, microscopes making use of electrons rather than light were designed. The electron is associated with a wave form that has a length equal to that of x rays and much shorter than that of visible light waves. X rays, however, cannot be focused easily because of their great energy, while electrons, which carry an electric charge, can easily be focused by magnetic fields. Such *electron microscopes* made objects visible that were far too small to study in ordinary microscopes.

While electron microscopes could focus on single giant molecules, single atoms remained beyond the horizon.

In 1936, the German physicist Erwin W. Mueller began work on a new principle whereby a very fine needle tip was made to emit electrons, or positively charged ions, in a vacuum. These electrons or ions would travel in a straight line to a fluorescent screen which would be lit at the points of impact.

In 1955, Mueller built the first *field-ion microscope* (in which ions were stripped off a needle tip by an electric field). The pattern that appeared on the fluorescent screen was the pattern of atom arrangement on the needle tip, magnified about five million times. In effect, individual atoms could be "seen" as dots in the pattern. So far, this works only for a few metals that have particularly high melting points, but even so, valuable information on atom arrangements, in perfect and imperfect crystal forms, has been obtained.

Game Theory

A GAME, in its usual sense, is some artificial activity designed to give pleasure. Usually, though, man's competitive instinct is such that the pleasure arises from pitting the individual's luck or skill against that of others and, of course, winning.

The game may involve pure chance, as in tossing coins; pure skill, as in chess; a mixture, as in poker; physical prowess, as in most athletic games. Even solitaire games are not without an adversary. The chance fall of the cards (that is, the randomness inherent in the universe) is, in this case, the adversary.

Games with known and limited rules, dealing with fixed numbers of pieces, limited areas and times of play, lend themselves to the mathematical analysis of *optimum strategy*, the steps that will insure the greatest chance of winning. To achieve the best result, one must assume the adversary is also using optimum strategy (but he has a different hand, different abilities, or differs perhaps only in that he moves second while you move first).

What applies to an ordinary game may also apply to the serious aspects of life. Business is a game played between competing producers and the consumer; war is a game between nations; even scientific research is a game between scientists and the universe.

The Hungarian mathematician John von Neumann applied mathematical analysis to the development of schemes of optimum strategy in these more serious games, based on the principles developed by dealing with games as simple as matching pennies. This originated the subtle mathematical treatment called *game theory*. He, with the economist Oskar Morgenstern, collaborated in writing a book, *The Theory of Games and Economic Behavior*, in 1944, and this, together with the development of computers, has brought game theory into prominence since World War II.

Gemini, Project

Once the one-man space flights of Project Mercury had been successfully completed in 1963, the National Aeronautics and Space Administration (NASA) went on to the next step. The attempt was now to be made to send up two men at a time. With two men in a larger capsule, it would be possible to arrange to have the vessel undergo maneuvers, change its orbit, attach itself to a second capsule that had been sent into orbit at another time, and so on.

The two-man launchings planned by NASA were referred to as *Project Gemini*. Since Gemini is the name of one of the constellations of the zodiac, one might think that there was some astrological significance to this but there wasn't.

The Latin word *geminus* means "twin," and twin brothers (or sisters) would be *gemini*. In the Greek myths, the most famous twins are Castor and Pollux (who were brothers of Helen of Troy). In the skies, there are two neighboring first-magnitude stars with no other bright stars very close. They looked like twins, so one was named Castor and the other Pollux. They retain those names to this day. Naturally, the dim stars in the neighborhood were thought of as representing the images of two young men, their arms linked, and the constellation was called *Gemini*.

Project Gemini, however, refers not to the constellation, but to the two men in the capsule who would be circling in the heavens like *The Heavenly Twins*, as the constellation Gemini is sometimes called in English.

Project Gemini proved successful. On March 16, 1966, one of the Gemini vessels joined an orbiting vessel while both were moving through space, the first *docking* in space. In 1965, an American astronaut left his vessel for a *space walk*, as a Soviet cosmonaut had done three months before.

Genetic Code

IN 1902, biologists decided that the chromosomes, small structures in the cell nucleus, controlled the inherited characteristics of the cell. Cell chemistry and, therefore, its characteristics are controlled by the many enzyme proteins present in the cell, each of which hastens a particular chemical reaction. Therefore the chromosomes must somehow direct the production of particular enzyme molecules.

The chromosome consists of protein and nucleic acid and it was taken for granted that it was the protein portion that was important. The chromosome protein might consist of models of the proteins making up the cellular enzymes. Each cell might manufacture enzymes according to the nature of its chromosome protein.

In 1944, however, the American biochemist Oswald T. Avery demonstrated that it was the nucleic acid that governed the synthesis of enzyme molecules. Biochemists were now faced with a riddle. How could the nucleic acid molecule, which was utterly different from protein, serve as a model for enzymes?

Investigation into nucleic acid structure moved into high gear. It turned out that nucleic acids were every bit as large and complicated in molecular form as proteins were, but nucleic acid molecules were built up of units called nucleotides, while proteins were built up of units called amino acids.

It further turned out that groups of three neighboring nucleotides represented a particular amino acid. A given nucleic acid, made up of a chain of such triplets, could serve as a model for the construction of a protein with a corresponding string of amino acids. But which group of nucleotide triplets corresponded to which amino acid? This correspondence was called the *genetic code* (from a Greek word meaning "birth," because you were born with a set of controlling chromosomes) and it was worked out in the 1960s.

Geochemistry

CHEMISTRY, the study of the composition of matter and the transformations it could be made to undergo, is a very old science that dates back into prehistory. When men learned to ferment grape juice, to make brick and glass, to smelt metals, they were being chemists.

With the passing of the centuries, men learned more and more about chemistry. Chemists studied various minerals, learned to form new gases, identified new elements, and evolved the atomic theory. It was not until the nineteenth century that the concept was advanced of the chemical structure of Earth as a whole and the transformations of the planetary chemistry in the course of Earth's evolution. The Swiss chemist Christian F. Schönbein coined the word *geochemistry* in 1838 for this branch of the science, the prefix "geo-" coming from the Greek word for "earth."

To gain real knowledge of the chemistry of Earth as a whole, there had to be a wide study of minerals from every region of Earth's crust. Methods had to be devised for studying atomic arrangements within the minerals and for developing theories as to which arrangements were most probable. Finally, methods were developed for determining something about the chemistry of Earth's deep interior.

It was not until the 1920s that planetwide crust sampling, the use of x-ray techniques, and the study of earthquake waves traveling through Earth's interior supplied the necessities. In the 1910s and 1920s, the science reached adulthood in Norway, particularly as a result of the work of the Swiss-born Norwegian chemist Victor M. Goldschmidt.

Since then, astronomical studies have brought in more and more information about the chemistry of the universe as a whole to give rise to the still newer subject called *cosmochemistry* (which means, obviously, "chemistry of the cosmos").

97

Geoid

EVEN THE ANCIENT GREEKS knew that Earth was a sphere but it was not until the time of Sir Isaac Newton in the 1680s that it was possible to show that it could not be a perfect sphere. Newton showed that because Earth rotated, it had to bulge outward in the equatorial regions. Careful measurements in the eighteenth century showed that Newton's theory was correct.

The diameter of Earth, measured from one point on the equator to an opposite point, is twenty-six miles greater than the diameter from pole to pole. Earth is an *oblate spheroid*, where *spheroid* comes from Greek words meaning "sphere-shaped." (If the diameter from pole to pole were the greater, the Earth would be a *prolate spheroid*. Both *oblate* and *prolate* come from Latin words meaning "to bring forward," as though either the equator or the poles were reaching out.)

Of course, Earth's surface is actually quite irregular, but even the highest mountain top is less than six miles above sea level, and even the deepest ocean bottom is only seven miles below sea level. What's more, mountainous regions are composed of fairly light rock, while the sea bottom is made up of dense rock, so that gravitational pull doesn't change as much as one would expect from the height alone.

In recent decades, geologists measured the actual pull of gravity at different places on Earth and calculated what the shape of Earth would be like if the actual surface were lowered or raised to make the pull of gravity equal everywhere on the surface. This equal-gravity shape is the *geoid* (Earth-shaped). Under ideal conditions, the geoid would coincide with a perfect oblate spheroid.

Actually, it turns out that there are small bumps and flattenings. In some places the geoid is as much as fifty meters farther from Earth's center than it ought to be if the planet were a perfect oblate spheroid, and in some places, fifty meters closer. This is because of uneven densities.

Gerontology

THROUGH MOST OF HUMAN HISTORY, life expectancy was short. Until the middle of the nineteenth century, the average life expectancy, even in the most prosperous and advanced regions, was about 35; and the general level of nutrition was such that men were usually distinctly old at 50. In the 1860s, however, the germ theory of disease was put forward by the French chemist Louis Pasteur, and methods were developed to combat a number of serious ailments that had till then killed millions of people each year.

This was followed by better notions of hygiene, by the discovery of vitamins in the early decades of the twentieth century, and by the discovery of antibiotics still later. Nowadays, the general life expectancy is 70 or more in the advanced regions of the world and has climbed rapidly in even the underdeveloped regions.

To be sure, even in the times when life was most insecure and short, there were always a few who lived on into extreme old age, but these were few indeed. In modern times, the increase in life expectancy has meant a rapid rise in the numbers of the aged, and this has brought special problems to medicine and sociology. Considerable research is now being conducted into the manner in which cells and tissues age, and this is called *gerontology* (from Greek words meaning "the study of old men"). The medical treatment of aged people generally is *geriatrics* (from Greek words meaning "medicine of old age").

Despite all the advances that have brought about increases in average life expectancy, the maximum life expectancy stays where it was, not much over a hundred years. But even if old men must still die, it is the aim of specialists in gerontology and geriatrics to make the last period of life as free of pain and discomfort as possible.

Gibberellin

JAPANESE RICE FARMERS had long been aware that every once in a while some rice plants would suddenly greatly elongate their stalks. They would grow so tall that their stalks would buckle and they would die. The Japanese farmers called this *bakanae*, which can be translated into English, rather charmingly, as "foolish seedling disease."

In 1926, a Japanese botanist, E. Kurosawa, was able to show that the seedlings that were so foolish as to overgrow and die were actually infested by a fungus, which (he felt) must be liberating a compound that accelerated growth. The fungus in question bore the scientific name of *Gibberella fujikuroi*. (The first part of the name is from a Latin word meaning "humpbacked," from the appearance of its cells. The second is from the name of the Japanese scientist who first described this particular species.) The substance which caused the growth was therefore named *gibberellin*.

In 1938, two closely related compounds were isolated from extracts of the fungus by three Japanese chemists. They were called *gibberellin A* and *gibberellin B*. (Partly because of the break in communications between Japan and the rest of the world during World War II, it was not until the 1950s that any work in the field was done outside Japan.)

It turns out that gibberellin is produced in higher plants, too, not just in fungi — something first established in 1956. Gibberellin is, indeed, a group of plant hormones, rather similar in effect to the auxins, but with more complicated molecules. It is merely the fact that the fungus produces too great a quantity of the hormone that makes it dangerous to the plants.

Gibberellins, in appropriate quantities, can be used to hasten plant growth to a nondangerous extent and thus increase crop yields.

Gigantopithecus

THROUGH THE 1920s and 1930s, archaeologists uncovered fragments of fossil bones in a cave about thirty-one miles southwest of Peking, China. These belonged to a primitive type of man with a brain distinctly smaller than that of modern man, but larger than that of any living ape. (All manlike species with brains larger than apes and with characteristics, such as upright posture, resembling those of modern man, are referred to as *hominids,* from a Latin word for "man").

Among the places in China where the search for hominid remains was intense were the native pharmacies. These sold powdered "dragon bones" (actually fossils) because they were thought to have medicinal properties. In 1935, a Dutch archaeologist, Gustav H. R. von Koenigswald, came across a large tooth in such a pharmacy, and in the next few years discovered three more. They looked very much like human teeth, but were unusually large. For a man to have teeth so large, he would have to be nine feet tall or so.

Speculation arose concerning the possibility of a prehistoric race of giant men. After all, the legends of many people speak of giants; even the Bible does.

World War II put an end to further investigation, but between 1957 and 1968, four jawbones were discovered into which teeth would fit that would be large enough to match von Koenigswald's find. The jawbones, however, were clearly apelike in nature. Apparently, there once existed a giant species of primate that was not manlike, but gorillalike — the largest primate who ever existed. It was about nine feet tall and its diet was very much like that of man, so that it developed similar teeth.

It was now named *Gigantopithecus* from Greek words meaning "giant ape." It probably was not the source of the legends about giants, though, for it has been extinct for at least a million years.

Global Village

ALTHOUGH the future of mankind seems dark in many ways, there are bright aspects. Proper education might teach the world's people how to live together in peace and how to take those measures needed to preserve the environment.

For education to reach all the world's billions some technological advance is needed, and a possible solution was pointed out in 1945 by the science-fiction writer Arthur C. Clarke. He was the first to explain the uses of communications satellites and the manner in which as few as three such satellites at a height of 22,000 miles above Earth's surface might provide world communication at the speed of light.

Since then, the dream has come to birth. In 1965, *Early Bird*, the first commercial communications satellite, was launched. In 1971, the more sophisticated *Intelsat IV* was launched with a capacity for 6000 voice circuits and 12 TV channels. In the future, the possible use of laser light in combination with communications satellites may make many millions of channels available for the world's population.

It seems within the realm of possibility that every person on Earth can be placed in potential touch with every other. Closed TV circuits might exist in such numbers that all business conferences can be held through images, with no one having to move physically from his place. Documents and books could be facsimiled and their images transported at once. Every person on Earth could be placed easily in touch with any source of information and with any event of cultural importance. There would be no backwaters, no hicks.

The entire planet Earth might then have the character of a village in which everyone knows everyone and where all information is totally available. Those who look forward to this kind of future speak, therefore, of the *global village*.

Gravitational Lens

ACCORDING TO ALBERT EINSTEIN's theory of relativity, light traveling through a gravitational field moves in a curved path. If a ray of light from a distant star passes very close to the surface of the sun on its way toward Earth, the light would curve slightly toward the sun's center and would seem to change its position compared to other stars farther from the sun.

Ordinarily we can't see stars close to the sun, but during a total eclipse we can. In 1919, astronomers observed a total eclipse to see if stars near the sun had changed their positions. They had, and the results were dramatic support for Einstein's theory, promulgated three years earlier.

Light also curves in passing from air to glass. When light passes through a lens, it will bend and meet at a point on the other side. Light rays gathered from a large area and concentrated in this fashion make the objects emitting or reflecting the light rays seem larger and brighter.

If one astronomical object is located directly behind another as viewed from Earth, the light from the farther object, in approaching us, will skim the surface of the nearer object on all sides. All the light from the farther object will bend slightly inward toward the center of the nearer object. The rays of light converge considerably by the time they reach us and the farther object will look larger and brighter than it really is. This effect is that of a *gravitational lens*.

Some astronomers wonder if the quasars, which seem so unusually bright, are affected by a gravitational-lens effect. Probably not, but in 1988 it is expected that one particular star will move in front of a more distant one and on that occasion the idea of the gravitational lens will be tested.

Graviton

THE FOUR FORCE FIELDS are (1) the strong nuclear interaction, (2) the electromagnetic interaction, (3) the weak nuclear interaction, and (4) the gravitational interaction. All make themselves felt across a vacuum, possibly through the continual exchange of particles. The electromagnetic interaction works through the constant interchange of photons, and the strong nuclear interaction through the constant interchange of pions.

Both photons and pions are known and have been studied. Are there exchange particles for the other two interactions also? In particular, what about the gravitational interaction? It is by far the weakest of the four interactions and if there is a particle associated with it, it must carry very little energy and be extremely hard to detect. Physicists have suggested that such a particle must exist. Using the "-on" ending common for subatomic particles since the discovery of the electron, they have named it the *graviton*.

All particles can be detected in the form of waves as well as of particles. Very energetic particles are much easier to detect in particle form, while very unenergetic particles are much easier to detect in wave form. Thus, while all masses are constantly giving off and absorbing gravitons, these might be detected most easily as *gravitational waves*.

Since 1957, the American physicist Joseph Weber has been trying to detect gravitational waves. He makes use of a pair of aluminum cylinders, 153 centimeters long and 66 centimeters wide, suspended by a wire in a vacuum chamber. A gravitational wave moving over it would distort the cylinders very slightly. Such a wave, coming from far out in space, would sweep over the entire planet Earth. Weber therefore places his two cylinders hundreds of miles apart. When both distort in precisely the same way at precisely the same time, he feels that a gravitational wave has passed by. In 1969, he reported a number of such events.

Greenhouse Effect

IT IS POSSIBLE to grow plants in large glass buildings, even in cold weather. This is because glass is transparent to the short waves that make up visible light but not to the longer infrared waves.

When sunlight strikes the glass buildings that house the plants, it passes through. Its energy is absorbed within and is reemitted in the form of infrared. Because the infrared cannot get out easily, it piles up, raising the temperature within the structure. Such a structure is a *greenhouse* because the plants within are green, even when the land is barren outside. The phenomenon whereby energy enters in one form and cannot escape in another is called the *greenhouse effect.*

In Earth's atmosphere, oxygen and nitrogen are transparent both to visible light and infrared. Carbon dioxide and water vapor, however, transmit only visible light. This means that sunlight can reach Earth's surface easily during the day, but at night, when Earth is reradiating infrared, it has difficulty escaping. Earth's surface is warmer than it would be, therefore, if there were no carbon dioxide or water vapor in the air.

The planet Venus has a very thick atmosphere that is mostly carbon dioxide. The greenhouse effect is enormous there and Venus' surface temperature is nearly 400° C. This was determined from radio-wave emission in 1956 and was confirmed when the Venus probe *Mariner II* passed near Venus in 1962.

The greenhouse effect may be of increasing importance on Earth. The burning of vast quantities of coal and oil since 1900, and especially since 1940, is slowly increasing the carbon dioxide content of the air. And even a small increase may raise Earth's temperature to the point where the ice caps will melt and raise the oceans' level so that millions of square miles of the continental lowlands will gradually be flooded.

Green Revolution

THE FIRST POPULATION EXPLOSION we know of in human history came with the development of agriculture some ten thousand years ago. Once plants were deliberately cultivated, a given tract of ground could support more people than could be supported by simply gathering or catching what food happened to be available.

Ever since then, world population has managed to expand because more and more land has been given over to agriculture, because fertilizers have increased crop yields, because pesticides have killed off insect pests that devour crops, and because the use of power machinery makes it possible to do the work of farming more efficiently.

Another method for improving the yield of the land is to concentrate on those strains of particular plants that grow more rapidly than others or that are more resistant to cold or to disease.

The last method was the course followed by the American agricultural scientist Norman Ernest Borlaug when he was sent to Mexico in 1944 to study methods for improving wheat production.

He developed new strains of dwarf wheat by crossing a Japanese variety with native Mexican wheat. The new strains were high-yield and resistant to disease, so that by 1960 Mexico's wheat production had increased tenfold and it had become an exporting nation.

Borlaug went on to develop varieties of wheat and rice that would flourish in the Middle East. The result was called the *Green Revolution,* for land turned greener, with denser, healthier growths of plant life. The immediate danger of famine receded.

Borlaug himself has pointed out, however, that the increase in food supply would serve no purpose if the human population continued to increase wildly; for then the danger of famine would return — and pose a worse threat than ever.

106

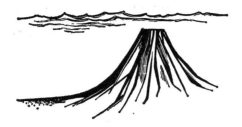

Guyot

UNTIL MODERN TIMES, virtually nothing was known about the sea bottom. It was covered by miles of sea water, and most people assumed that there was just a flat, featureless plane under all that water. In the 1870s, however, the vessel *Challenger* was the first to study the ocean bottom carefully. By dropping lines overboard, the scientists on board found that the Atlantic Ocean was shallower in its mid-region than on either side.

Detailed studies could not be made, however, until the technique of listening to sound echoes was developed in the 1920s. Once that was introduced, geologists began to make out the outlines of a vast mountain chain, larger and higher than anything on dry land, running the length of the Atlantic Ocean and into the Pacific and Indian oceans. The *Mid-Oceanic Ridge* virtually encircles Earth and its highest peaks emerge through the ocean surface to form islands such as the Azores.

In 1942, the American geologist Harry H. Hess, a naval officer, had echo devices installed on his ships, and while he crossed and recrossed the Pacific Ocean on war business, he also studied variations in ocean depth. In this way, he discovered isolated mountains studding the sea bottom, mountains whose tops were flat, as though they had been worn away by wave action, but which were a couple of thousand feet below the surface where wave action is nil. They may have been mountains which were originally above the surface till the sea bottom subsided or the sea level rose.

These submerged mountains, discovered by the hundreds, are called *seamounts* or *tablemounts*, but their most dramatic name was given to them by Hess. He called them *guyots* (pronounced "gee-oze'," with a hard *g*) in honor of an old professor of his at Princeton, a Swiss-born geographer, Arnold H. Guyot, who founded modern physical geography.

107

Hadron

THERE ARE FOUR TYPES of force fields known to man, each of which makes attraction (and sometimes repulsion) felt across a distance. The first to be studied was the gravitational interaction, which is by far the weakest of the four but is associated with such huge conglomerations of matter, like the sun and Earth, that the total effect is very large.

A second is the eletromagnetic, which is associated with the electrically charged particles that make up the atom, the electron and the proton chiefly. Atoms and molecules hang together because of electromagnetic interactions. All the pushes and pulls we associate with living beings and with mechanical devices (except for gravitational effects) are electromagnetic.

It was not until the middle 1930s that any other force fields were studied. These arose out of the necessity of explaining the atomic nucleus, which held together despite the fact that it contained protons and neutrons only. The protons were all positively charged and should have repelled each other strongly. Yet they stayed together. It seemed there must be a nuclear interaction that held them together (with the help of the uncharged protons) against this repulsion.

In order to account for what went on within the nucleus, two kinds of interaction were required. There was the *strong nuclear interaction*, which was a hundred thirty times as strong as the electromagnetic, and the *weak nuclear interaction*, which was far weaker than the electromagnetic but was stronger than the gravitational.

Only the more massive subatomic particles, such as the proton and neutron, were capable of responding to strong nuclear interactions. These massive particles (including some mesons and the very massive hyperons) were grouped together as *hadrons*, from a Greek word meaning "strong."

Half-Life

In the 1890s, it was discovered that certain atoms constantly give off particles from their nuclei and change into different types of atoms. They are said to undergo *radioactive breakdown.*

If every radioactive atom of a particular kind had a fixed life — that is, if each existed a particular length of time before breaking down — then a group of identical atoms would exist unchanged for a while and then suddenly break down all at once, loosing vast energies.

That doesn't happen. Instead, a large collection of identical radioactive atoms are continually giving off small quantities of energy, as though a few atoms were breaking down during each small interval of time. Some might break down today, some tomorrow, some a billion years from now. The time when a particular atom breaks down would seem to be entirely a matter of chance. There is no use, therefore, in speaking of the lifetime of a radioactive atom. That can have any value.

In any large collection of such atoms, though, there is a particular probability that within a certain length of time a certain fraction of the total number will have broken down. We can tell exactly when one-tenth the total number will have broken down even if we can't tell which tenth. (Insurance companies can predict how many American males will die in the next year though they couldn't tell which particular ones would die.)

It proved particularly convenient to select a period of time in which exactly half the radioactive atoms in any conglomeration would break down. This period of time is called the *half-life.* Thus, the half-life of uranium 238 is 4.5 billion years, while the half-life of its sister variety, uranium 235, is only 0.7 billion years. Some half-lives are much shorter. The half-life of radium 226 is 1620 years and that of polonium 212 is 0.0000003 seconds.

Hallucinogens

THE BRAIN, like every other part of the body, performs its functions through certain chemical reactions. These are produced by stimuli brought to the brain through the senses. It is possible to change the brain chemistry by taking into the body substances that interfere with these chemical reactions. In that case, the body will respond to stimuli that don't relate to the outside world. Objects seem to be sensed that are not really there; while other objects which are really there may be ignored. The results are *hallucinations*, from a Latin word meaning "to wander in the mind."

Certain plants contain chemicals which can produce hallucinations. The peyote cactus and a mushroom called *Amanita muscaria* contain such chemicals. Sometimes these plants are eaten in primitive religious celebrations because the hallucinations are thought to be glimpses of another world (or an escape from this one). Another substance that produces hallucinations is hashish, one form of which is marijuana.

In 1943, a Swiss chemist, Albert Hofmann, was studying an organic compound called *lysergic acid diethylamide* and accidentally got a few tiny crystals of it on his fingers. He happened to touch his fingers to his lips and was soon overcome by odd hallucinations. It took him a full day to regain normality. He began careful studies and found the chemical could always produce hallucinations in very small doses. The name was soon reduced to an abbreviation of the three words. Since the German word for "acid" is *Säure* and Hofmann spoke German, the abbreviation was *LSD*.

Since many young people foolishly began to play games with their minds by taking LSD and other such substances, hallucination-producing drugs became important to study. They are now lumped together under the general name *hallucinogens* (producers of hallucinations).

Holography

IN ORDINARY PHOTOGRAPHY, a beam of ordinary light, reflected from an object, falls on a sensitized film. Wherever light falls, the film darkens, forming a *negative*. From the negative a *positive* is formed, a flat two-dimensional representation.

Suppose instead that a beam of light is split in two. One part strikes an object and is reflected with all the irregularities that this would impose on it. The second part is reflected from a mirror with no irregularities. The two parts meet at the photographic film, and the interference between the two light beams is recorded. The film that records this interference pattern seems to be blank, but if light is made to pass through, it takes on the interference characteristics and produces a three-dimensional image, which can then be photographed in the ordinary manner from different angles.

The notion was first worked out by the Hungarian-British physicist Dennis Gabor in 1947, and he called it *holography*, from Greek words meaning "whole writing." The whole of the image, not part, was recorded.

Gabor's idea could not be made practical with ordinary light in which wavelengths of all sizes moved in all directions. The interference produced by two such beams of light would then be too chaotic to give sharp images.

The introduction of the laser changed everything. Laser beams had uniform wavelengths all moving in the same direction. In 1965, Emmett N. Leith and Juris Upatnieks, at the University of Michigan, were able to use laser light to produce the first holograms. Since then, the technique has been sharpened to the point where holography in color has become possible, and where the photographed interference-fringes produced with laser light could then be viewed with ordinary light.

Hydroponics

Since the invention of agriculture, about ten thousand years ago, the tilling of the soil has been the most important method by which human beings have assured themselves of a food supply. *Agriculture* is from Latin words meaning "to cultivate a field." A similar word from Greek roots would be *geoponics* (meaning "to till the ground").

People always assumed that plant life was nourished by the soil, and indeed some soil was fertile and some barren; some yielded good harvests, unless poorly watered, and some did not, even when well watered. In the nineteenth century, however, it came to be realized that plants made use of small quantities of mineral substances in the soil, by dissolving those substances in water. The main nourishment came from the air while the soil itself (aside from the small quantities of useful minerals) served merely to hold the plant.

It therefore seemed possible to grow plants in a water solution of those substances necessary to its nutrition. In that way, the nutritive substances would be particularly available to the plant and their proportions would be easily controlled. Plants ought then to grow larger and healthier despite the absence of solid soil. This would be *hydroponics* (to till the water).

Hydroponics requires a highly developed chemical industry for support, and as long as ordinary agriculture is possible it is unlikely that hydroponics will ever replace it. Hydroponics can merely be used as supplement. In one respect, though, hydroponics may soon come into its own. When men take off on long space voyages beyond the moon, it would be impractical to carry with them enough food to last for years. It would surely be better to grow food of one sort or another, and on board the spaceship the techniques of hydroponics would be most suitable. Astronauts will then be expert hydroponicists, or water farmers, as well.

Hyperon

ALTHOUGH THE ELECTRON and the proton have electric charges of identical size, the proton is 1836 times as massive as the electron. In 1930, the neutron was discovered. It has no electric charge, but is even a tiny bit more massive than the proton.

For nearly twenty years, the proton and neutron were the most massive single particles known and it was suspected they might be the most massive possible. In 1947, however, two English physicists, George D. Rochester and Clifford C. Butler, detected a V-shaped track in a cloud chamber, and found that one of the branches of the track was that of a particle about 2200 times the mass of an electron and therefore about a fifth again as massive as a proton or neutron.

Since the capital form for the Greek letter *lambda* is shaped like an upside-down *V*, Rochester and Butler, in honor of the shape of the track that revealed the new particle, called it a *lambda particle*.

In the years that followed, other particles, even more massive, were detected and studied. All were unstable, enduring for not more than a ten billionth of a second before breaking down to protons, neutrons, and smaller particles.

Each of the different groups of massive particles was given a Greek letter name, following the lambda precedent. There was a group of *sigma particles* with a mass 2360 times that of an electron; a group of *xi particles* with a mass of about 2600 times that of an electron; and in 1964, an *omega particle* was discovered that was 3300 times as massive as an electron; nearly twice as massive as a proton.

All these massive particles were grouped together under the name of *hyperons*, from the Greek word *hyper*, meaning "above" or "beyond." The masses of the hyperons were, you see, above or beyond those of the proton and the neutron.

Imaginary Number

A NUMBER MULTIPLIED by itself yields a *square number* and the original number is, itself, the *square root* of that square number. Thus, the square of 5 is 5 $\times$ 5, or 25. This means that 25 is the square of 5 and 5 is the square root of 25. A square root doesn't have to be an integer; in fact, it usually isn't, but is an endless decimal instead. The square root of 2 is 1.414214 . . . and so on, endlessly.

Now consider numbers with signs before them, positive and negative numbers. Two positive numbers multiply to yield a positive product, and two negative numbers do so likewise. This means that $+5 \times +5 = +25$ and $-5 \times -5 = +25$. The square root of $+25$ is both $+5$ and -5.

But what about the square root of a negative number, such as -25? There isn't any in the ordinary system, since neither positive numbers nor negative numbers will do. You can write any negative number, $-x$, as $+x \times -1$. The square root is then the square root of $+x$ times the square root of -1. The whole problem boils down to the square root of -1.

The square root of -1 was called an *imaginary number*, because it didn't exist in the ordinary number system. In 1777, the Swiss mathematician Leonhard Euler symbolized the square root of -1 as i for *imaginary*.

Imaginary numbers are not, however, imaginary. Consider a horizontal line with a zero point located on it. Points to the right can represent positive numbers, points to the left negative ones. Another line, up and down through zero, can represent positive and negative imaginary numbers. The entire plane is defined by a combination of the two numbers called *complex numbers*. Complex numbers are indispensable in the mathematics of electrical engineering, for instance. And in the 1960s, physicists talked of tachyons with masses expressed in imaginary numbers. Ordinary or *real numbers* are thus no more real than the imaginaries.

Infrared Giant

In 1905, the Danish astronomer Ejnar Hertzsprung considered the fact that some red stars, like Antares and Betelgeuse, were very bright, and some, like Barnard's Star, were very dim. Both were equally low in surface temperature, for all were no more than red-hot; warmer stars glow yellow, white, and even blue white. The only way that a cool star that gives off only a dim red glow from each portion of its surface can nevertheless appear bright is for it to have a large surface.

Hertzsprung therefore suggested that there were *red giants* and *red dwarfs* among the stars. There seemed to be no red stars of intermediate size and that began the line of reasoning that developed modern notions of the way in which stars evolved with time.

To begin with, a large cloud of dust and gas slowly condenses, getting hotter and brighter as it does so, for a million years or more, until it finally reaches a stable situation in which it can remain for billions of years. In the process of condensing, when it glows red-hot, it will pass through a red-giant phase.

After it has consumed its chief nuclear fuel (hydrogen) through billions of years of shining, a star will begin fusing helium and still more complicated atoms. When it does so, it expands enormously and, in the process, its surface cools and again it passes through a red-giant phase.

In condensing originally, a star passes through a phase when it radiates chiefly infrared light. Again, when a dying star expands, it may expand so much that its surface radiates chiefly infrared. In 1965, astronomers at Mt. Wilson Observatory used a special telescope with a large plastic mirror to scan the sky for spots rich in infrared. Within a couple of years, they found thousands of objects that were barely shining and those were the *infrared giants*.

Interferon

ANTIBIOTICS have helped the medical profession control many bacterial infections. That is because a particular chemical can affect bacterial cells more seriously than it will affect host cells. With proper dosage, bacteria can be stopped with no harm to the host. Viruses are not so easily handled. They do their work inside the cell they parasitize, using the very chemical machinery of the cell. To kill one by outside chemicals means killing the other.

Nevertheless, virus diseases don't always kill. In most cases, the affected organism recovers; sometimes it is hardly affected. Thus the organism must have natural defenses. The organism forms protein molecules capable of reacting with the virus and preventing it from harming the cells. These antibodies sometimes persist through life, so that anyone who recovers from measles, chicken pox, mumps, and certain other virus diseases is immune to them thereafter.

In the 1930s, there began to be signs that cells could fight off viruses even without producing antibodies. There was *virus interference.* The fact that viruses invaded some cells made it harder for them to invade other cells in the organism. In 1957, the English biochemist Alick Isaacs showed that the entry of viruses into cells stimulated the formation of tiny quantities of a small protein that left the cell, entered other unaffected cells, and guarded them against penetration. This limited the effects of the virus infection. The protein which interfered with the virus Isaacs called *interferon.*

Each species produces an interferon that works only on its own cells. You can't infect chickens, isolate their interferon, and hope to help humans with it. However, attempts are being made to learn how to stimulate the production of interferon in humans by the addition of some harmless substance that will then leave the body immune to particular viruses thereafter.

Ion Drive

A ROCKET makes its way out of the atmosphere and through space by means of jet propulsion. A propellant burns, sending a jet of exhaust gas in one direction, so that the rest of the rocket is pushed in the opposite direction. The propellant is usually a chemical fuel that burns in some active substance like oxygen, so that this method of accelerating a rocket may be called a *chemical drive*.

A chemical drive can lift many tons of matter into space against the pull of Earth's gravity. Chemical fuels are, however, rapidly consumed. Many tons of propellant are required to launch a rocket. Once out in space, where the gravitational pull of distant worlds is comparatively small and there is no air resistance, it takes only short bursts of fuel exhaust to correct orbits, but even they use up fuel quickly.

Is there any kind of propellant that would last longer? One suggestion involves ions, which are atoms that have lost one or more electrons and which therefore carry a positive electric charge. These ions can be repelled by another positive charge and forced out the rear of a rocket engine, at an enormous speed.

The ions, moving backward, will push the rocket ship forward. They will do so only very slightly because the ions are so small, so such an *ion drive* would not work in the neighborhood of Earth's surface. Only far out in space where there is no necessity to overcome air resistance and no strong gravitational field will the ion drive manage to make its very slight push felt.

However, the ions produced by a large supply of matter will last for a very long time. There would be only a tiny push, but it would be kept up for years and that tiny push — push — push would eventually bring the rocket to speeds close to that of light. It may be that only by some sort of ion drive can we ever hope to reach the distant stars.

117

Jansky

KARL JANSKY was a radio engineer employed by Bell Telephone Laboratories. In 1931, they set him the task of studying the causes of static, which constantly interfered with radio reception and with radiotelephony from ship to shore. Static has a number of causes, from thunderstorms and aircraft to nearby electrical equipment.

Jansky, however, detected a new kind of weak static from a source which, at first, he could not identify. It came from overhead and moved steadily. At first, it seemed to be moving with the sun. It gained slightly on the sun, however, to the extent of four minutes a day. This is just the amount by which the vault of the stars gains on the sun. Consequently, the source seemed to lie beyond the solar system.

By the spring of 1932, Jansky had decided the source was in the constellation of Sagittarius, the direction in which the center of the Galaxy was located.

He published his findings in December 1932, and this represented the birth of *radio astronomy*. It was the first indication that there were vast sources of radio-wave energy in the universe. Astronomers learned to receive and interpret microwaves (the shortest radio waves) in addition to light waves. Microwaves penetrate dust clouds that light waves cannot, so that *radio telescopes* could receive information about objects totally invisible to ordinary optical telescopes. Furthermore, very distant or unusual objects, which could not be made out by ordinary telescopes, revealed themselves through radio waves.

Jansky himself did not continue to work with radio astronomy, for he was more interested in his engineering. He died in 1950 of a heart ailment while still only 45, so that he lived to see only the infancy of the science he had founded. He is, however, not forgotten. The unit of strength of radio-wave emission is now called the *jansky* in his honor.

Janus

ONCE THE TELESCOPE was invented, it was soon found that planets had small companion bodies circling them. In 1610, the Italian scientist Galileo Galilei discovered four satellites circling Jupiter.

Other such discoveries were made, with the larger satellites of the closer planets detected first. In 1892, the American astronomer Edward E. Barnard discovered Amalthea, a fifth satellite of Jupiter, far smaller and also closer to its planet than the other four. It was the last satellite to be discovered by eye.

By 1950, the resources of photography had raised the number of satellites to thirty-one. Earth had one, Mars two, Jupiter twelve, Saturn nine, Uranus five, and Neptune two. Astronomers were sure that further satellites existed but felt they were probably too small or too distant to detect except by space probes.

Saturn offered an interesting problem, though. In addition to its nine satellites, it had a brightly shining set of rings. The innermost known satellite was Mimas, but if there were any still closer, it might well be lost in the glare of the rings. Every fourteen years, however, Saturn and Earth are so situated that the rings of the former are presented edge-on to the latter. The rings are so thin that they then cannot be seen.

The year 1966 was one of those in which the rings of Saturn were not seen from Earth, and the French astronomer Audouin Dollfus then carefully investigated the close neighborhood of Saturn. His search was rewarded. Only 4000 miles outside the outer edge of the ring system there was a satellite of respectable size (300 miles in diameter). It would certainly have been discovered long ago were it not for the rings. As the tenth and latest of the satellites of Saturn to be discovered and the first in order counting out from the planet, he called it *Janus* after the Roman god with two faces, one looking forward and one backward. What better name to call a satellite that was both last and first?

119

Jet Plane

ONE OF THE FUNDAMENTAL LAWS of the universe is the law of conservation of momentum. This says that if an object at rest is to begin to move in a particular direction, something else must move in the opposite direction. Thus, when we walk our feet push Earth in the opposite direction, but only unimaginably slightly, because momentum is velocity times mass, and Earth's mass is much greater than our own.

This holds true for all means of propulsion. Automobile tires push Earth in the opposite direction. Oars, paddlewheels, screw propellers, all push water backward so that a ship can go forward. Airplane propellers push air backward so the plane can go forward.

There is a limit to how fast airplanes with propellers can go, because there is a limit to how fast propellers can turn without flying apart. In that case, why burn fuel in order to turn a motor, in order to turn a propeller, in order to throw air backward? The burning fuel turns into hot gases that exert great pressure in expanding. If you give those gases an opening behind, they will expand through that opening and be themselves thrown backward. The backward movement will force the airplane forward, and we can go straight from burning fuel to the desired motion without going through all the other steps.

The hot exhaust gases, which, so to speak, throw themselves through the opening behind, are a *jet* (from the French word for "to throw"). A plane which pushes forward because of the jet pushing backward is a *jet plane*. In 1939, an Englishman, Frank Whittle, flew a reasonably practical jet plane, and the invention was perfected during World War II.

Since the war, jet planes have come into commercial use and world travel has been revolutionized as large planes carry great numbers of passengers farther and faster than would have been possible in the days of propeller planes.

Jet Stream

MEN HAVE ALWAYS BEEN AWARE of moving currents of air, the winds. Naturally, they have known them best where they could directly experience them — on or near the surface of Earth.

After the airplane was invented, new kinds of winds were experienced. During World War II, high-flying American planes, making their way westward toward the war zones near Asia, were astonished to find themselves making poor time and using up a surprising amount of fuel. They were fighting a headwind, apparently, and a very powerful one.

When the phenomenon was studied, it turned out that there was a steady, very strong, west-to-east wind blowing at a height of eight miles or so above the surface of Earth. Its speed sometimes reached up to 200 miles an hour. Because this stream of air moved with the force of a jet of fluid emerging from a narrow nozzle, it was called the *jet stream*. Actually, it turned out, there were two jet streams, located between the mainly cold arctic mass of air and the mainly warm tropic mass of air, north and south of the equator. The northern jet stream circles Earth in the general latitude of the United States, the Mediterranean Sea, and northern China. The southern jet stream moves around Earth in the general latitude of Argentina, Chile, and New Zealand.

The jet streams meander, often curving into eddies far north or south of their usual course. Airplanes now take advantage of the opportunity to ride on the swift winds. More important still is their influence on moving air masses nearer Earth's surface. These control the surface weather so that studying the variations in the jet-stream course makes it possible to be a little surer about the long-range prospects in weather forecasting.

So far, the southern jet stream has been far less studied than the northern one, but it is likely that they are not very different.

121

Juvenile Hormone

INSECTS HAVE a tough cuticle made of *chitin* (from a Greek word for "a coat of armor"). While the insect is growing, the cuticle presents a problem. The chitin of a growing insect larva must periodically split. The larva wriggles out of the split cuticle and quickly grows a new and larger one. This is called *molting*. After a certain number of molts, the larva undergoes more radical changes to become an adult insect.

The process of molting seems to be an automatic one controlled by the production of an insect hormone called *ecdysone*, from a Greek word meaning "to molt." But then what stops the molting and suddenly causes the larva to undergo the other changes on the way to adulthood? Apparently, a second hormone must be involved.

In 1936, an English biologist, Vincent B. Wigglesworth, cut off the head of a certain species of insect to see what would happen in the absence of any hormones that might ordinarily be produced in the head. The insect lived (a head is not as important to an insect as to a vertebrate) but at once began the changes leading to adulthood.

Apparently the head produced a hormone that kept the insect a larva and molting. When no head hormone existed, the changes toward adulthood began. Since the head hormone acted to keep the insect in the immature state, it was called *juvenile hormone*.

The American biochemist Carroll Williams began to study the juvenile hormone in the 1950s. He found that juvenile hormone, placed on insects that had already begun the change toward adulthood, was absorbed through the skin. It then acted to stop the change and the insect died.

In short, the juvenile hormone, or any synthetic compound that worked similarly, could possibly be a very useful kind of insecticide. It works only on insects and on no other organism, and usually on only one group of insects, not others. It would thus lack the disadvantages of other insecticides.

Kaon

AT FIRST, physicists were aware of only a few subatomic particles. Gradually, they learned to work with larger energies and to make use of more delicate detection devices. As a result, they learned to produce, identify, and study many additional particles.

In 1944, French physicists noted a cloud-chamber track in their study of cosmic rays that indicated a particle about 1000 times as massive as an electron and therefore half as massive as a proton. Such tracks were found again, usually in association with a particle called a pion. The unknown particle and the pion were apparently formed together by the impact of cosmic ray particles on atoms. The two particles, after formation, would then move off in different directions, leaving a V-shaped track. This was therefore called a *V-event* and the heavy particle was called a *V-particle*.

Eventually, though, it turned out that V-events were fairly common and that not all of them involved this particular particle. A new name therefore had to be found. Since the mass was intermediate between that of an electron and a proton, it belonged to the meson family. To distinguish it from other mesons, it was called the *K-meson* and this name is frequently shortened to *kaon*.

The kaon is very unstable and only endures for about a hundred millionth of a second before breaking up in any of six different ways to form (usually) smaller mesons.

The different ways in which kaons break up proved important, for it turned out that a kaon could break down in such a way that a certain subatomic property called parity could be either odd or even. Until then, it had been thought that parity had to stay odd, or had to stay even in any particle change. The finding that it could be either odd or even in this case led to an interesting change in physical theory.

123

Karyotype

EVERY CELL NUCLEUS contains a group of chromosomes that can be arranged in pairs. The chromosomes resemble a mess of stubby strands of spaghetti all tangled up and it is extremely difficult to count the number per cell. For instance, it was long thought that human cells each contained forty-eight chromosomes, in twenty-four pairs. It was not until 1956 that a very painstaking count showed the true number to be forty-six, in twenty-three pairs.

Fortunately, this problem no longer exists. A technique has been devised whereby treatment with a low-concentration salt solution in the proper manner swells the cells and disperses the chromosomes. They can then be photographed and that photograph can be cut into sections, each containing a separate chromosome. If these chromosomes are matched into pairs and then arranged in the order of decreasing length, the result is a *karyotype* (from Greek words meaning "picture of the nucleus").

The karyotype offers a new tool in medical diagnosis, for it shows clearly whether there are any missing chromosomes, extra chromosomes, or damaged chromosomes. Such imperfections, by adding or subtracting or distorting a whole series of genes (which control the chemistry of the cell), can produce serious birth disorders.

There is, for instance, a disease called *Down's syndrome*, because it was first described in 1866 by the English physician John L. H. Down. Babies born with it are mentally retarded and have physical deficiencies as well. (The disease is sometimes called *mongolism* or *mongolian idiocy* because one of the symptoms is an eyelid shape that looks like those common in East Asia. This is a poor name, though, since the disease has nothing to do with East Asians.)

In 1959, three French geneticists, Jerome J. Lejeune, M. Gautier, and P. Turpin, found that cell nuclei in Down's-syndrome patients had forty-seven chromosomes, not forty-six. There was an extra chromosome at position 21.

K-Capture

ATOMIC NUCLEI are only stable when they possess protons (with a positive electric charge) and neutrons (with no electric charge) in a certain ratio. When there are too many neutrons, one of those neutrons must change into a proton. Suppose a neutron has no charge because it possesses equal amounts of both positive and negative. It can become a proton by getting rid of the negative charge. For that reason, an atomic nucleus with too many neutrons can become stable by emitting the negative charge in the form of an electron.

In similar fashion, an atomic nucleus which is unstable because it has too many protons can convert a proton to a neutron by getting rid of a positive charge. Such nuclei become stable by emitting positrons (electronlike particles that carry a positive electric charge).

In 1936, the Japanese physicist Hideki Yukawa showed that it was theoretically possible for a nucleus to capture one of the electrons in the outer reaches of the atom. Capturing a negative charge was equivalent to emitting a positive one; therefore nuclei that become stable by emitting a positron might capture an electron instead. The American physicist Luis W. Alvarez detected actual cases of this in 1938.

The electrons are grouped outside the nucleus in a series of shells, up to seven in number in very complicated atoms. This was found to be so, originally, because under certain circumstances atoms emit x rays at various levels of frequency, one for each electron shell. The x-ray frequencies are labeled in alphabetical order, starting arbitrarily with K, so that there were $K, L, M \ldots$ frequencies in order of decreasing energy. The electron shells were correspondingly labeled $K, L, M \ldots$ in order of increasing distance from the nucleus.

The K-electrons (those in the K-shell) are nearest the nucleus and are most apt to be captured. The phenomenon of electron capture is frequently called *K-capture* for that reason.

Krebs Cycle

THE CHIEF BODY FUEL is a six-carbon sugar called *glucose*, which is always present in the blood stream and is picked up by every cell in the quantities needed. A little energy can be obtained by breaking glucose molecules into two-carbon fragments. Fat and protein molecules can also be broken down to two-carbon fragments.

The bulk of the energy available to living tissue then arises from the combination of the two-carbon fragments with oxygen to form carbon dioxide and water. Naturally, biochemists were very interested in discovering the details of this process.

One way of doing this was to chop up muscle tissue and measure the rate at which it would take up oxygen. When the rate of oxygen uptake falls off, one or another organic molecule is added. Those which cause the oxygen uptake to jump again must play a part in the oxidation chain. In 1935, the Hungarian biochemist Albert Szent-Györgyi showed that four closely related four-carbon molecules were involved. The German biochemist Hans A. Krebs added additional compounds to the list and by 1940 was able to complete the chain.

The two-carbon fragment was added to a four-carbon compound, with the formation of a six-carbon compound as result. This six-carbon compound (*citric acid*, because it occurred in citrus fruits) was gradually broken down to the four-carbon compound by steps that formed carbon dioxide and water and liberated energy. The four-carbon compound was then ready to pick up another two-carbon fragment.

The reactions moved in a cycle from four-carbon to six-carbon and back to four-carbon, grinding up a two-carbon fragment at each turn. This is sometimes called the *citric acid cycle* because it begins with the formation of citric acid. It is more often called the *Krebs cycle* after the man who worked it out.

126

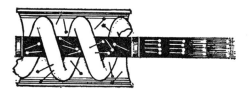

Laser

In 1953, the American physicist Charles H. Townes designed a device that would absorb a weak beam of microwaves and emit a strong beam of exactly the same sort (provided it had the proper energy supply). This process was *microwave amplification by stimulated emission of radiation* or, by taking the initials of the important words, a *maser*.

Why should only microwaves be amplified in this way? The same system could be used for other wavelengths, for those of visible light, for instance. In 1960, the American physicist Theodore H. Maiman used a bar of synthetic ruby for the purpose. Its molecules were energized to a high-energy level and then, when a feeble beam of red light of the proper wavelength was allowed to fall upon it, an intense beam of the same wavelength of red light emerged.

This was *light amplification by stimulated emission of radiation* or a *laser*. (It might also be called an *optical maser* but the one-word version has been universally adopted.)

The light emerging from a laser consists of waves that are all traveling in exactly the same direction and are, so to speak, all moving up and down in unison. In ordinary light of all other kinds, waves are moving in all sorts of ways, possess all sorts of energies, and move up and down without relationship to each other. The laser light does not, therefore, spread out as ordinary light does, but remains a tight beam which spreads out only slightly even if it travels all the way to the moon. This light beam, sticking together as it seems to do, is called *coherent light*.

Because the light beam sticks so tightly together, a great deal of energy is concentrated in a small area, and at the point of impact a laser beam can produce temperatures far higher than that of the surface of the sun.

Lawrencium

BEGINNING IN 1940, physicists learned to manufacture elements with atoms more complicated than those of uranium, with its atomic number of 92. By 1960, ten of these elements, from 93 to 102, were formed. One way of forming them was to bombard the atoms of elements already produced with small atomic nuclei. There might be coalescence.

In 1961, for instance, an American team under Albert Ghiorso bombarded californium (98) with nuclei of boron (4). When the two nuclei coalesced, they occasionally underwent radioactive changes and these produced a few atoms of element 103. These atoms were unstable and half of them had already broken down within eight seconds, but while they existed, they were a new element.

Previous elements that had been synthesized had been named after deceased scientists whose work had significance in nuclear science. Element 99 is *einsteinium* after Albert Einstein, who first showed that mass could be converted to energy; 100 is *fermium* after Enrico Fermi, who first bombarded uranium with neutrons; 101 is *mendelevium* after Dmitri I. Mendeleev, who first developed the periodic table of elements. Following this precedent, 103 was named *lawrencium* after Ernest O. Lawrence, who invented the cyclotron.

In 1965, a Soviet team under Georgy N. Flerov bombarded plutonium (94) with nuclei of neon (10) and obtained element 104, which they named *kurchatovium* after Igor V. Kurchatov, who led the group that developed the Soviet nuclear bomb. The Soviet method of forming 104 has not been confirmed elsewhere but the element has been formed in other ways. Americans call it *rutherfordium* after Ernest Rutherford, who carried through the first man-made nuclear reactions. In 1970, Ghiorso and his team produced 105, which they named *hahnium* after Otto Hahn, who first discovered uranium fission.

Lepton

IN THE 1860s, it was discovered that electricity could be forced through a vacuum, producing a new type of radiation. This was called *cathode rays*, because the radiation seemed to emerge from that part of the electric circuit called the cathode. For a generation these cathode rays were studied and were eventually found to be made up of particles carrying a negative electric charge.

It turned out that atoms could be made to carry a negative electric charge and that atoms of different mass could carry exactly the same charge. Indeed, the electric charge seemed to come in packets of fixed size, all of which were exact multiples of a certain minimum. If atoms of different size carried this minimum electric charge, a given mass of small atoms would carry more than the same mass of large atoms.

In 1897, the English physicist Joseph J. Thomson found that a given mass of cathode rays carried many times as much electric charge as the same mass of even the smallest atoms (those of hydrogen). The conclusion was that the individual cathode ray particles were much less massive than even the smallest atom. These particles were named *electrons* and turned out to be $\frac{1}{1837}$ as massive as a hydrogen atom.

The electrons are far smaller than the other particles making up atoms of matter. No one knows why they are so much smaller than the other two, the proton and the neutron. There are other particles, not commonly found in ordinary matter, that are as small or smaller. In fact, the neutrino seems to be a particle with no mass at all.

The electron and certain other particles are subject to weaker forces than those involved in the considerably more massive particles and react more slowly. These particles, subject to weak forces, have in recent years come to be grouped together as *leptons*, from the Greek *leptos*, meaning "weak."

Linac

FROM 1930 ONWARD, physicists have constructed devices to accelerate subatomic particles, making them gain more and more energy. The more energetic they are, the more effectively they will smash into atomic nuclei and the more they will tell us about nuclear structure.

Some devices use a strong magnetic field to whirl the particles in a circle. Another method is to use an alternating electric field to push a particle in a straight line through a series of tubes in each of which the electric field changes its direction as the particle enters. As the particle speeds up, with each fresh push in each fresh section of tube, the section has to be made longer than the one before. The great length of such a device is a disadvantage when compared with the much more compact method of whirling the particles in a circle. Then, too, it proved difficult to keep the electric field alternating in exact rhythm with the particles passing from one tube to the next.

As a result the *linear accelerator* (or *linac* for short), first tried in 1931, passed into disuse. And yet it had its advantages, too. When particles move in a circle, the necessity of constantly changing the direction in which they travel uses up energy which might otherwise go into speeding them up. In the case of massive particles like protons, this can be borne; but for light particles like electrons, this energy loss sets a sharp limit to the total energy they can receive.

In a linear accelerator, electrons are not forced to lose energy through direction change. Once techniques were worked out to keep the electric field alternating with greater precision, it became desirable to build a large linac for producing electrons with record energies. Stanford University has now constructed a linear accelerator for the purpose which is two miles long.

130

Mach Number

Ernst Mach was an Austrian physicist whose most important work was in the philosophy of science. In the nineteenth century, scientists were self-confident indeed, and charged forward in their search for greater knowledge without worrying too much over the fine points of the methods of reasoning which they were using. By and large, they blundered on correctly, but Mach began to insist they stop to analyze what they were doing. His views were unsettling and unpopular and he did not have much influence in his own time. He did, however, impress Albert Einstein, who went on to revolutionize science with his theory of relativity (which Mach in his last years, ironically enough, refused to accept).

One line of practical investigation initiated by Mach consisted of experiments in airflow, which he published in 1887. He was the first to take note of the sudden change in the nature of the airflow over a moving object as it reaches the speed of sound.

Sound travels through any medium by means of the natural motion, back and forth, of the atoms or molecules in that medium. When a material object forces its way through the medium, the molecules get out of the way by means of their natural swing. As the material object approaches the speed of sound, it is approaching the speed of the natural motion of the molecules of the medium. Those molecules can no longer get out of the way and their behavior changes.

After World War II, airplanes began to move at speeds approaching that of sound and the behavior of the airflow became of great interest. The ratio of the velocity of the planet to the velocity of sound in air, at a given temperature and density, was called the *Mach number*. If the planet moved at exactly the speed of sound, it was moving at *Mach 1*; if it was moving one and a half times the speed of sound, that was *Mach 1.5*, and so on. (The speed of light, the fastest speed possible, is about Mach 900,000.)

Macromolecule

AFTER THE ENGLISH CHEMIST John Dalton had advanced the atomic theory in 1803, it was soon found that each substance was made up of a combination of atoms that held together firmly to make up a *molecule* (from Latin words meaning "a small mass").

The first molecules studied consisted of only a few atoms each. Water molecules, for instance, consisted of three atoms each, two hydrogens and one oxygen (H_2O). Common salt could be considered as made up of combinations of one atom of sodium and one of chlorine ($NaCl$). Sand molecules are made up of three atoms (SiO_2) while other common substances of the earth's rocks, limestone ($CaCO_3$), alumina (Al_2O_3), and hematite (Fe_2O_3), are made up of five each.

These are all molecules characteristic of inanimate nature and are therefore *inorganic* (from Latin words meaning "not life"). The situation is different in the case of *organic* substances, however, those that are characteristic of matter that is living, or that was once living.

Organic molecules are combinations of greater numbers of atoms, but of a smaller variety of different atoms. Most organic molecules consist of carbon, hydrogen, and oxygen (plus nitrogen or a scattering of others sometimes). Acetic acid molecules are made up of eight atoms ($C_2H_4O_2$); citric acid molecules of twenty-one atoms ($C_6H_8O_7$); table sugar molecules of forty-five atoms ($C_{12}H_{22}O_{11}$); and so on.

That was only the beginning. The substances most characteristic of living tissue, and most important, are giant molecules made up not merely of dozens, but of hundreds or thousands or even millions of atoms. These *macromolecules* ("macro-" is from a Greek word meaning "large") include such substances as the proteins and nucleic acids which are the keys to life, so that the study of macromolecules is really the study of life.

132

Maffei 1

IN THE DISTANT REACHES of the universe, there are galaxies by the millions. The individual galaxies do not usually exist singly, but tend to group into *galactic clusters*. As many as ten thousand galaxies may cling together under mutual gravitational influence.

Is our own Milky Way Galaxy also part of a cluster? Two cloudy patches visible in the sky of the Southern Hemisphere are called the *Magellanic clouds* (because they were first described in 1521 by the chronicler accompanying Ferdinand Magellan on the expedition that first circumnavigated the globe). These are two galaxies, much smaller than our own and quite nearby. They might almost be considered satellite galaxies of ours.

In addition, there is the Andromeda Galaxy (so named from the constellation in which it is to be found) which is probably even larger than our own. It is the closest of all the large galaxies, not much more than two million light-years away. At distances between those of the Magellanic clouds and the Andromeda there are about twenty other galaxies, all small ones. All these form the *Local Group*, of which we and Andromeda seemed to be the only two giant members.

But not all parts of the sky can be clearly seen. In the zone of the Milky Way itself, there is so much dust that the sky is obscured. In 1968, an Italian astronomer, Paolo Maffei, was searching the sky and found patches of infrared in one of the dustiest regions of the Milky Way. (Infrared will penetrate dust somewhat more efficiently than visible light will.) Studying the regions closely, he discovered two galaxies with only one per cent of their light filtering through the dust. They were close enough to be part of the Local Group and large enough to be giants. They have been named *Maffei 1* and *Maffei 2* in honor of the discoverer and are our new (and unsuspected) neighbors.

Magic Number

In 1916, it was found that the electrons in an atom were distributed in groups that enclosed the nucleus like spherical shells. As one progressed outward from the nucleus, the shells were larger and could hold more electrons. The innermost shell could hold only two electrons, the next eight, the next eighteen, and so on. Each shell was divided into subshells and it was found that the chemical properties of elements depended on the distribution of the electrons in their atoms among the shells and subshells. When certain subshells were filled, for instance, an atom was particularly inert, and did not engage in chemical reactions.

In the early 1930s, when it was found that the atomic nucleus was made up of protons and neutrons, physicists began to wonder at once if those particles were arranged in shells as electrons were. It took a great deal of energy to probe the nucleus, however, so the matter could not be worked out easily.

One way of considering the question was to take note whether any atomic nuclei were particularly stable. That might indicate their protons or neutrons had filled certain shells or subshells. It turned out that when a nucleus contained 2, 8, 20, 50, 82, or 126 of either protons or neutrons, it was particularly stable. In 1949, the German physicist J. Hans D. Jensen called these *magic numbers*. Later, feeling perhaps that this was an overdramatic name ill-suited to science, he used the term *shell numbers* instead, but the earlier term remains more popular.

In 1949, Jensen and (independently) the German-American physicist Marie Goeppert-Mayer worked out a system for the arrangement of shells within the nucleus, based on the magic numbers. It is more complicated than the electron shells and not as well established, but it is a beginning.

Magnetohydrodynamics

THE STUDY of the flow of liquids such as water under the impulse of various kinds of pressure is called *hydrodynamics* from Greek words meaning "motion of water." Gases can flow, too, but do so in a more complicated fashion because gases can easily compress and expand, whereas liquids cannot. Since both liquids and gases flow, they are grouped together as *fluids* and the study of their flow is *fluid dynamics*. However, it has become customary to use the term *hydrodynamics* as a synonym for fluid dynamics, and to treat gas flow as well as liquid flow under that title.

If gases are heated to thousands of degrees, the atoms that make them up are broken down into electrically charged particles, and these will interact with a magnetic field. Under such circumstances, the magnetic field may exert pressure on them and cause them to flow. The study of such flow is *magnetohydrodynamics*.

The control of hot gas by a magnetic field is an essential part of attempts to achieve controlled fusion, so magnetohydrodynamics may be a key to the energy sources of the future. It may also be a key to the more immediate energy sources of today.

At present, fuel must be burned; the heat must boil water to steam; the steam must turn a wheel; the turning wheel must generate electricity. At each stage, there is loss of energy, and in the end, the electricity formed represents only thirty to forty per cent of the energy present in the fuel to begin with.

It is possible to use the fuel to heat gases to high temperatures of five thousand degrees and send them through a magnetic field to generate a current directly. This magnetohydrodynamic method could conceivably raise efficiency to some sixty per cent and effectively double the energy we can obtain from our ordinary fuels of today.

Magnetosphere

In 1600, the English physician William Gilbert showed, from the behavior of the magnetic compass, that the planet Earth must be a huge spherical magnet. Like any other magnet, Earth has magnetic poles, a *north magnetic pole* in northern Canada and a *south magnetic pole* on the rim of Antarctica. Between the two lie imaginary *magnetic lines of force*, curving outward from one pole to another and being at their highest point above Earth's surface halfway between the poles. These lines represent the direction along which magnetic attractions and repulsions are directed.

In 1957, a Greek amateur scientist, Nicholas Christofilos, advanced a theory according to which charged particles in the neighborhood of Earth would be trapped by the planet's magnetic field and would spiral back and forth from one magnetic pole to the other along the lines of force (the *Christofilos effect*, as it is now known). Near the poles the magnetic lines of force approach Earth's surface and the charged particles there interact with the molecules in the upper atmosphere, producing the aurora.

Christofilos' work was ignored at first, but in 1958, a region of charged particles enveloping Earth were discovered by rocket investigations and named the *Van Allen belts*, after James A. Van Allen, the scientist in charge. The particles in the Van Allen belts turned out to behave as Christofilos had predicted, and he is now in the United States working as a professional scientist.

The Van Allen belts were soon renamed. The various layers about Earth have names ending in "-sphere" because of their spherical shape. *Atmosphere* (sphere of vapor) is itself an example. Since the belts are involved with Earth's magnetic field, the Van Allen belts are now called the *magnetosphere*.

 # Mare Cognitum

In 1609, the Italian scientist Galileo Galilei first looked at the moon with a telescope and saw mountains and craters. There were also extensive regions which seemed to lack mountains and craters.

As the seventeenth century progressed, other astronomers began to draw maps of the moon and to give names to its chief features. The flat regions without mountains and craters they called *maria* (a Latin word for "seas"). Some may have thought they were really regions of water. Actually, it was soon recognized that the moon lacked an atmosphere and on an airless world, open stretches of water are not at all likely. The flat regions are seas only in the sense that they are fairly flat seas of compacted dust.

Nevertheless, they not only retained the name of maria, but individual stretches kept the romantic names originally given them, such as *Mare Humorum* (Sea of Fluid), *Mare Imbrium* (Sea of Rain), *Mare Undarum* (Sea of Waves), *Mare Spumans* (Sea of Foam), and so on. Some names, to be sure, are better fitting. *Mare Serenitatis* and *Mare Tranquillitatus* are "Sea of Serenity" and "Sea of Tranquillity," and the surfaces of those dead seas are indeed serene and tranquil.

In the 1960s, men began to get a closer look at the moon's surface than even the best telescope could afford. In 1964, the United States sent lunar probes directly toward the moon. These headed for a crash landing, taking photographs and sending them back to Earth as they approached. On July 28, 1964, *Ranger VII* approached the surface of the moon, sending back 4316 photographs, the last few from a distance just feet away. It struck near the northwest edge of Mare Nubium (Sea of Nubia) in the midst of a small flat stretch which was quickly named *Mare Cognitum* (Sea of the Known) because, thanks to *Ranger VII*, it was better known than any other part of the moon — at that moment.

137

Mariner Program

THE FIRST OBJECTS sent into space in 1957 and 1958 were artificial satellites that circled Earth in as little a time as ninety minutes. Even satellites sent into orbits around Earth that stretched hundreds of miles away completed those orbits in a matter of hours.

It is different for objects sent into space at speeds greater than Earth's escape velocity. They do not circle Earth, but recede indefinitely and go into an orbit about the sun. Such orbits are not completed for many months.

The first *space probes* (objects sent away from Earth toward other worlds), launched in 1959 and 1960, were aimed at the moon. Some passed around the moon and back toward Earth; some passed on and disappeared into the emptiness of space. In either case, scientists were interested only in the reports of conditions in the neighborhood of Earth and the moon.

It takes only three days for a probe to travel from Earth to the moon and this is not a long journey. Men have made the round trip safely a number of times since then.

In 1962, however, space probes were being sent toward the planet Venus. Venus is, at its closest, twenty-five million miles from Earth, a hundred times as far away as the moon is. A *Venus probe* takes about four months to reach Venus. The thought of that long, lonely voyage through space was very impressive. The probe was like Samuel T. Coleridge's Ancient Mariner, who was "Alone, alone, all, all alone;/ Alone on a wide, wide sea."

The planetary probes were therefore part of the *Mariner program.* *Mariner II* was launched in August 1962 and successfully passed Venus in December. In November 1964, *Mariner IV* made the even longer voyage to Mars, which it successfully passed in July 1965.

Mascon

THE INTENSITY of the gravitational field increases with mass, so that gravitation shapes the form of large bodies. These become spheres under the pull of their own gravity. They can have unevennesses on the surface, so that mountains and valleys are present, but these are very small compared to the whole body. A quickly rotating body will bulge at the equator, but a slowly rotating one like the moon is an almost perfect sphere. Gravity also makes the density of a body vary from center to surface in the same way in all directions. This means the gravitational field should be the same at a given distance from the body's center in all directions. It was assumed this was true of the moon.

Between 1966 and 1968, a number of satellites were placed into orbit about the moon (and were hence called *Lunar Orbiters*). Assuming an even distribution of gravity about the moon, those vessels were expected to follow certain paths. They deviated slightly from those paths, however, and it appeared that over certain sections of the moon's surface, its gravitational field was a trifle stronger than over others. The field was stronger over the flat seas than over the mountains and craters, and it had to be assumed then that the density of the moon's crust was higher in the sea regions than in the mountainous areas.

There had to be concentrations of mass under those seas (the only reasonable explanation) and the term *mass concentrations* was quickly abbreviated to *mascons*. One possible reason for the mascons is that the seas were formed by the collision of large, dense meteors with the moon, and that those meteors (iron, perhaps) still lie buried under the seas. Or else the seas contained water once and the dry surface is now covered by thick and dense layers of sedimentary rock.

Perhaps with further lunar explorations, we will find out which, if either, of these explanations is correct.

Maser

THE NEW KNOWLEDGE concerning atomic structure, which came with the opening of the twentieth century, made it clear that a particular atom or molecule could absorb a photon containing a fixed quantity of energy and enter a higher-energy state. In 1917, Albert Einstein showed that if a photon of just the right size struck an atom or molecule that was already in the higher-energy state, it would drop to the lower, emitting a second photon just the size of the first and traveling in the same direction.

If all the atoms or molecules in a particular volume of matter were in the higher-energy state, a photon of the right size would bring down one molecule. Then there would be two photons, each of which would bring down one molecule, producing four photons, and so on, until in a very short time, a gigantic flood of photons would be emerging, all moving in the same direction.

In 1953, the American physicist Charles H. Townes devised a method for pumping a quantity of ammonia molecules into a higher-level state. When he then allowed a very weak beam of microwave photons, representing the proper energy difference between the high-energy and low-energy levels of the molecule, all the molecules were dropped to the low-energy level, and a large flood of microwaves resulted.

By stimulating the emission of radiation through the original weak beam, that beam was amplified into a large one. This was *microwave amplification by stimulated emission of radiation*. Taking the initial letter of the key words of this phrase, we have *maser*.

Since Townes's first maser, the Dutch-American physicist Nicolaas Bloembergen has devised a system with three levels so that pumping and emission can go on at different photon sizes. Emission was then steady, and the first *continuous maser* was invented.

140

Mercury, Project

AFTER THE SOVIET UNION launched its first artificial satellite, *Sputnik I*, the United States began to make plans at top speed for the exploration of space. While unmanned satellites are important, the American government was sure that it would be necessary to send a man into space eventually. After all, no machine yet invented can take the place of the inquiring human mind. The National Aeronautics and Space Administration (NASA) was established in 1958 to oversee this.

The first men in space would merely orbit Earth. To do this, however, the rocket carrying them would have to travel at a speed of five miles per second, and at that rate Earth would be circumnavigated in ninety minutes. (When Magellan's men made the first circumnavigation of Earth in 1522, it took them three years.)

The speed at which men in orbit could go around the planet was reminiscent of the winged messenger of the gods in the Greek myths, who moved so quickly in his journeys that he was pictured with wings on his sandals and cap. He was Hermes, and the Romans called him Mercury. The plan to put men in orbit was therefore called *Project Mercury*.

In 1961, two Americans were sent up in rockets and returned safely, but they did not go into orbit. On February 20, 1962, however, John H. Glenn, Jr., was sent into orbit and circled Earth three times before landing safely. (Two Soviet cosmonauts had been placed into orbit before that.)

Others followed Glenn, until on May 15–16, L. Gordon Cooper, Jr., was placed in orbit and circled Earth no less than twenty-two times, remaining in space for thirty-four hours before coming back safely. Project Mercury had proved a resounding success and had brought American achievements in this area abreast of those of the Soviets. The United States was then ready to go on to the more ambitious program of Project Gemini.

Meson

IN THE 1920s, the only subatomic particles known were protons and electrons and it was supposed that the atomic nucleus had some of each. Protons repel each other, but protons attract electrons, so the electrons were thought to serve in the nucleus as a kind of cement, holding it together.

In 1930, the neutron was discovered and it was soon realized that the nucleus had to be made up of protons and neutrons only. But what kept the protons from repelling each other, then, and breaking up the nucleus?

A Japanese physicist, Hideki Yukawa, suspected a hitherto unknown nuclear force might exist which held the protons and neutrons together. In 1935, he worked out the requirement to make an attractive force strong enough to hold the nucleus together and yet be felt only over the width of the nucleus.

It turned out that certain particles had to be endlessly and rapidly exchanged between the protons and neutrons of the nucleus to make such a force work. Yukawa calculated they would be about 270 times as massive as electrons, and therefore about $\frac{1}{7}$ as massive as a proton or neutron. No such particles, intermediate in size between electrons and protons, were known.

In 1936, however, an American physicist, Carl D. Anderson, who was studying the tracks of cosmic rays in cloud chambers, came across tracks that had to be formed by particles of such intermediate mass. He labeled the particle *mesotron* ("meso-" coming from a Greek word meaning "intermediate"). This was eventually shortened to *meson*.

A number of different kinds of particles in this mass range were eventually discovered, and the term *meson* has been used for the entire group of particles that are more massive than electrons but less massive than protons and neutrons.

142

 # Messenger-RNA

BEGINNING IN THE 1940s, biochemists could demonstrate that the nucleic acid molecules of the chromosomes guided the synthesis of specific enzyme molecules. Yet the nucleic acids of the chromosomes were confined to the nucleus, deep within the cell, while enzyme molecules were manufactured in the cytoplasm, outside the nucleus. Somehow, the information concerning the structure of the nucleic acid molecule had to be carried out to the cytoplasm. There had to be a messenger.

The nucleic acid of chromosomes is a particular variety called *deoxyribonucleic acid*, which is frequently abbreviated *DNA*. This is not, however, the only kind of nucleic acid in the cell. Another kind, just as complicated as DNA and made up of very similar units, is *ribonucleic acid*, or *RNA*. Both have as a portion of their structure a sugar called *ribose*, but in the case of DNA there is an oxygen atom missing in the ribose portion, hence the "deoxy-" prefix.

RNA differs from DNA in that the former is located both in the nucleus and in the cytoplasm. Was it possible, then, that there might be a particular RNA molecule which could be synthesized in the nucleus, using one of the DNA molecules in the chromosomes as its model? This RNA molecule, carrying a copy of the structure of a particular DNA, could then move out into the cytoplasm where it could serve as a guide for the synthesis of an enzyme molecules. Such a RNA molecule would carry information from nucleus to cytoplasm. It would be the sought-for messenger and was almost inevitably called *Messenger-RNA*.

Messenger-RNA was first identified in bacteria, but in 1962, two American biochemists, Alfred E. Mirsky and Vincent G. Allfrey, demonstrated its presence in mammalian cells. It is now recognized as a universal part of the mechanism of the inheritance of physical characteristics from cell to cell and from generation to generation.

143

Metalliding

THERE ARE DOZENS of different metals that exist, but most of them, in pure form, are of only limited use. Gold, for instance, would seem perfect for jewelry, but pure gold is so soft it would lose its shape and wear away too easily. Add a little copper, though, and it is hard enough. Again, pure copper is too soft to use in tools and weapons or for knives and spears. Add tin and you get bronze, which is hard enough. Iron will rust, but the addition of some other metals will make *stainless steel,* a nonrusting metal that is mostly iron.

Mixing metals to form *alloys* will produce a myriad useful substances. The simplest method is simply to melt the two metals and stir them together. The alloy that results is a mixture all the way through. Another way is to plate a thin layer of one metal onto the surface of a bulky piece of another. This can be done mechanically or by means of an electric current which will layer a metal out of solution (electroplating). In that way, a small quantity of gold layered over another, much cheaper metal will make the whole object as beautiful and as resistant to rust as though it were solid gold.

In 1968, the American metallurgist Newell C. Cook tried to plate a silicon layer on a platinum surface, but found that under the conditions he used, silicon atoms worked their way below the surface. Instead of plating one substance onto another, he had prepared a metal with an outer skin that was an alloy, a situation more stable and permanent than plating. Cook called the process *metalliding.*

This process shows promise of economy. For instance, copper alloyed with two to four per cent of beryllium becomes extraordinarily tough. This can be achieved if copper is *beryllided* so that only the outer skin is alloyed. The entire piece of copper is just about as tough as though it were all alloyed but much less beryllium need be used.

Microwaves

VISIBLE LIGHT with the longest waves is seen by us as red. Light with still longer waves is invisible and is called *infrared* ("infra-" coming from a Latin word meaning "below"). The wavelength of infrared light is anywhere up to one millimeter.

In 1887, the German physicist Heinrich Rudolf Hertz first detected radiation of extremely long wavelength, far longer than infrared, generated from the spark of an induction coil. They were called *Hertzian waves* by some and *radio waves* by others. The latter name, which merely means "waves that radiate" (which all waves do) is quite meaningless, but it won out.

The term *radio waves* is sometimes applied to the entire range of waves from one millimeter to many kilometers. At first, it was the longer radio waves that proved most useful. They were reflected by layers of charged particles in the upper atmosphere so that they would bounce between Earth and sky and therefore follow the curvature of Earth for long distances. It was this which made long-distance radio communication possible.

On the other hand, the very short radio waves went right through the charged layers, and so not much attention was paid them. In the 1930s and afterward, however, they became important. They could penetrate clouds and fog and bounce off obstacles much better than longer radio waves could. The short radio waves were therefore used to detect incoming airplanes during World War II (radar). Then it was discovered that short radio waves were emitted by many astronomical bodies and *radio astronomy* was developed.

The short radio waves soon received a name of their own, *microwaves* ("micro-" coming from a Greek word meaning "small"). Microwave wavelengths are from one to sixteen millimeters.

145

Mid-Oceanic Ridge

IN 1853, when the Atlantic Cable was being laid, it seemed to those who were making soundings that there was a plateau in mid ocean. The Atlantic Ocean was shallower in the middle than at either end.

After World War I, soundings were made by listening to ultrasonic echoes. This gave much more detail of the rise and fall of the bottom than one could get by dropping a line. By 1925, it seemed there was a vast undersea mountain range winding down the Atlantic's center.

Later soundings elsewhere showed that the mountain range was not confined to the Atlantic. At its southern end, it curves around Africa and moves up the western Indian Ocean to Arabia. In mid-Indian Ocean, it branches, so that the range continues south of Australia and New Zealand and then works northward in a vast circle all around the Pacific Ocean.

At first the mountain range had been called the *Mid-Atlantic Ridge*, but when its world character was understood, it came to be called the *Mid-Oceanic Ridge*.

After World War II, the details of the ocean floor were probed with renewed energy by the American geologists Maurice Ewing and Bruce C. Heezen. In 1953, they discovered that a deep canyon ran the length of the ridge and right along its center. This was eventually found to exist in all portions of the Mid-Oceanic Ridge, so that sometimes it is called the *Great Global Rift*. There are places where the rift comes quite close to land. It runs up the Red Sea between Africa and Arabia and it skims the borders of the Pacific through the Gulf of California and up the coast of the state of California.

Through the rift, hot molten rock from below pushes upward and spreads the sea floor apart. This moves great plates of Earth's crust this way and that and is the cause of continental drift.

Mitochondria

IN THE 1830s, it was found that all living tissue was composed of tiny *cells* too small to be seen by the unaided eye, and each walled off from the others by a membrane.

It was supposed at first that each cell was a microscopic drop of homogeneous living fluid which, beginning in 1846, was called *protoplasm*. As microscopes improved, however, it became clear that the contents of the cells included dimly seen granules.

It was hard to see the details within the cell, because everything in it was more or less transparent. In the 1850s, however, chemists began to synthesize organic molecules, some of which were colored. Using these *synthetic dyes*, biologists found that some would be absorbed by some parts of the cell contents only. This meant that these portions of the cell contents would stand out against the rest in bright color.

Unfortunately, the dyes killed the cells and sometimes coagulated the clear substances within and created apparent granules that would not have existed in the dye's absence. Thus when a German biologist, Richard Altmann, reported certain granules in the outer regions of the cells, the observations were met with considerable skepticism. It was not till another German biologist, Carl Benda, repeated the observation in 1897 that scientists were convinced. The objects were called different names by different observers, but Benda had proposed the name *mitochondrion* (*mitochondria* is the plural form) and that was finally accepted.

The name isn't very descriptive. The granules, as observed, are like tiny threads or filaments and *mitochondria* is from Greek words meaning "filament granules." In the 1950s and 1960s, electron microscope studies showed the mitochondria to have a complex structure. Biochemists found it to be the powerhouse of the cell, containing enzyme systems that supervised reactions that liberated energy for the use of the body.

Mohole

THE STUDY OF EARTHQUAKE WAVES has shown that at the center of Earth there is a sphere of molten nickel-iron. Around this *nickel-iron core* is a layer of rocky substances called the *mantle*. The core and the mantle bear about the same position and proportions as the yolk and white of an egg.

In 1909, a Croatian geologist, Andrija Mohorovicic, was studying the manner in which earthquake waves traveled through the upper layers of Earth. To account for the time it took them to reach various points on the surface, he had to suppose that they bent sharply about twenty miles beneath the surface. This meant that the mineral structure above that level was distinctly different from that below. It was a *discontinuity* and came to be known, in his honor, as the *Mohorovicic discontinuity*, or the *Moho discontinuity* for short.

Above the discontinuity is Earth's *crust* (like the eggshell lying outside the egg white). It is of a different kind of rock than the mantle. It is possible that the crust was built up in the course of Earth's evolution and that the top of the mantle represents the planet's original surface.

It would be useful to penetrate through the crust, therefore, to the top of the mantle. The Moho discontinuity lies farther below the surface on dry land than at sea. At sea, the Moho discontinuity may be only eight miles below the surface and most of that is just water. In places, only three miles of solid crust would have to be penetrated.

Plans were made in the 1960s to drill down to the Moho discontinuity, and the drilling was slangily referred to as a *Mohole*. The prospect aroused considerable excitement among geologists, but it meant the expenditure of a lot of money which the government decided to spend in other ways. For the time, therefore, the Mohole has been shelved.

Molecular Biology

BIOLOGY (from Greek words meaning "study of life") is the science of living things. The word was first used in 1802 by the German naturalist Ludolf C. Treviranus. At first, biology concerned itself only with organisms and parts of organisms large enough to be seen with the unaided eye. In the seventeenth century (long before the word *biology*, but not the study, was invented) microscopes came into use and biologists could see what had hitherto been invisible.

It was discovered that all life forms were made up of tiny units, invisible to the unaided eye, which were called *cells*. In the nineteenth century, the inner structure of the cell was slowly worked out. Colored chemicals were used, for instance, because some of these adhered to some structures within the cell but not to others. The colored structures could then be studied in some detail.

Chemists went farther and began to study the molecules of those substances that characteristically made up cells. By the mid nineteenth century, this science of *biochemistry* was flourishing.

As scientific techniques grew subtler, chemists were able to study the complex molecules in greater detail. By the 1940s, for instance, methods were developed to break down large molecules into fragments that could be analyzed and then put back together again. In this way, the fine structure of large molecules could be worked out. By this method and by studies of the behavior of large molecules in beams of x rays, in electric fields, and under strong centrifugal effects, new understanding of life was grasped, with still more promised.

Biology came to deal more and more with the properties of the very large molecules of protein and nucleic acid. Since World War II, the term *molecular biology*, first used by the English biochemist W. T. Astbury, has come into fashion to describe this new study.

Monopole, Magnetic

ABOUT 1870, the Scottish mathematician James Clerk Maxwell worked out a detailed theory which related electricity and magnetism in a manner so intimately connected that you could not have one without the other. As a result, we speak of an *electromagnetic field*. When such a field fluctuates, radiation is given off. Light is an example of such *electromagnetic radiation*.

Maxwell's theory could be improved if electricity and magnetism were analogous in every possible way. There remains one great difference, however. An object may have one of two kinds of electric charge, positive or negative, but both need not be present on a single particle. An electron carries *only* a negative charge; a proton, *only* a positive charge.

There are also two kinds of magnetic charge, concentrated at the *north pole* and the *south pole*. (They are so called because the most impressive magnet known in early modern times was Earth itself and its magnetic charge was concentrated in the polar regions.) Every body which possesses a magnetic field, however, always has a concentration of each kind — *both* a north pole and a south pole.

In 1931, the English physicist Paul A. M. Dirac suggested that to make electricity and magnetism completely analogous, there ought to be particles containing only a north pole or only a south pole. These particles would be *magnetic monopoles*.

Physicists have been trying to detect monopoles ever since, but without success. If they exist, they would have to possess huge energies and could only be formed by use of such energies.

In 1969, the American physicist Julian Schwinger suggested that certain hypothetical electrically charged particles called quarks are also magnetic monopoles. He has suggested *dyon* as a name for this combination (with the prefix "dy-" from a Greek word meaning "two").

Mössbauer Effect

CERTAIN ATOMS emit gamma rays. In theory, these should be of a certain energy but as the gamma ray emerges, the atom emitting it recoils. The recoil takes up a certain amount of energy. When a number of atoms emit the gamma ray, each may recoil by a slightly different amount so that gamma rays of a fairly broad range of energies will be emitted.

In 1958, a German physicist, Rudolf L. Mössbauer, showed that when atoms are part of a crystal, the recoil is sometimes spread out over the entire crystal. There is then hardly any recoil at all since the crystal is so much more massive than a single atom. The gamma ray that issues is of sharply defined energy. This is the *Mössbauer effect*.

Gamma rays of the energy emitted by a particular crystal will be absorbed by another crystal of the same kind. (The gamma ray emitted by a particular crystal exactly fits it own nuclear structure, so to speak. If you make a key that fits a particular lock, you can then use it to open another lock identical to the first.) Gamma rays of a different energy will not be absorbed by the crystal.

An ordinary gamma ray beam, with a range of energies, cannot be dealt with sharply. Another crystal will absorb some of it and not the rest. A sharp Mössbauer-effect beam will, however, be absorbed entirely or not at all, a much more noticeable thing.

This has been used to check Einstein's general theory of relativity. According to Einstein's theory, a beam of gamma rays should lose energy very slightly if it must climb against the pull of gravity — or increase very slightly if it drops with gravity. If a beam of gamma rays, which a crystal can absorb, is allowed to drop from top floor to basement, the crystal will no longer absorb it at the basement. The energy has increased — very slightly, to be sure, but enough to be detected by the crystal — and Einstein's theory is supported.

Muon

THROUGH the early twentieth century, subatomic particles were best studied by the tracks they left passing through cloud chambers. Particles with electric charge knocked electrons out of the atoms they struck and left a trail of charged atoms (ions) behind. The chamber was saturated with water vapor and tiny water droplets formed around each ion. In this way the track of the subatomic particle was made visible.

The nature of the particle could be determined from the manner in which the track curved in the presence of a magnetic field. Particles with a positive charge curved in one direction; those with a negative charge, in the other. Massive particles carrying a unit electric charge curved gently; light particles carrying the same charge curved sharply. An experienced physicist could easily read off the tracks at sight.

In 1935, the American physicist Carl D. Anderson was studying cosmic rays. These were composed of energetic particles that struck atoms in the atmosphere and produced showers of subatomic particles of all kinds. Anderson was astonished to find that some tracks curved more sharply than those of protons but less sharply than those of electrons. They were particles of intermediate mass, the existence of which had been predicted a year earlier by the Japanese physicist Hideki Yukawa.

Anderson named the new particles *mesotrons* (from the Greek *meso*, meaning "intermediate"), and this was shortened to *meson*. As time went on, though, it turned out that Anderson's particle was by no means the only kind of object of intermediate mass. Other kinds of mesons were discovered and all had to be distinguished from one another.

Since Anderson's meson was the first to be found, it was given the honor of the initial letter of the word (*m*) as as its special marker, but the letter was used in its Greek analog, *mu*. The particle was therefore called the *mu-meson* and this has now been shortened to *muon*.

Mutagen

ABOUT THE TURN of the century, biologists noted that young were sometimes born without resembling either parent in a particular characteristic. This was termed a *mutation* by the Dutch botanist Hugo de Vries, from a Latin word for "change."

In order to learn about the mechanism of heredity, biologists studied these mutations and the way they were inherited. Noticeable mutations are usually quite rare, however, so methods were sought for increasing their frequency.

The American biologist Hermann J. Muller found that raising fruit flies at higher temperatures increased the incidence of mutations, but only slightly. It occurred to him to try x rays. An energetic x ray might hit a chromosome in some key spot introducing a change in the hereditary mechanism. By 1926, Muller had succeeded. X rays produced many mutations and they were now easy to study.

In 1937, the American botanist Albert F. Blakeslee discovered that the alkaloid *colchicine* interfered with the process of cell division in plants, so that cells with abnormal numbers of chromosomes were produced. These were the first mutations to be produced by chemicals.

During World War II, it was found that mustard gas somehow reacted chemically with the substances in chromosomes, altered them, and produced mutations. The search was on, and a rather long list of other chemicals capable of doing the same was discovered. These chemicals came to be called *mutagens*, the Greek ending making the word mean "capable of producing mutations."

It would appear that cancer results from some particular mutations (that in some cases may be virus induced). A mutagen may therefore bring about cancer. In that case, it is also a *carcinogen*, from Greek words meaning "to produce cancer."

153

Nanometer

In 1795, the French revolutionaries established a new system of measurement, so convenient and so logical that it has now been accepted (or is being accepted) by the whole world except for a few nations such as Nigeria, Ceylon, Gambia, Liberia, Jamaica — and the United States.

Essentially, the system establishes a particular unit of measurement, the *meter*, which is a unit of length equal to about 39.3 inches. Other units of the same sort are then indicated by different prefixes. A *kilometer* is a thousand meters, the prefix "kilo-" coming from the Greek word for "thousand." A *millimeter* is one thousandth of a meter, the prefix "milli-" coming from the Latin word for "thousand."

The original designers of this *metric system* did not make up prefixes for less than a thousandth, because at that time they scarcely seemed to be needed. Since then, however, scientists have been probing smaller and smaller measurements. They began to speak of *micrometers*, for instance, where a micrometer is a thousandth of a millimeter, or a millionth of a meter. The prefix "micro-" is from the Greek word *mikros*, meaning "small."

That was not enough, either. In 1958, the prefix "micro-" was made official and two prefixes, representing even smaller quantities, were established. A thousandth of a micrometer is a *nanometer* (from the Greek *nanos*, meaning "dwarf") and a thousandth of a nanometer is a *picometer*. There is no Greek word related to "pico-," but it may have originated by association with *picayune*, meaning "very small" or unimportant.

In 1962, it was established that a thousandth of a picometer is a *femtometer* and a thousandth of a femtometer is an *attometer*. Neither "femto-" nor "atto-" is related to any Greek word. All these new prefixes can also be used for units for measurements other than length.

 # Nerve Gas

POISON GAS was first used in warfare on a large scale during World War I. In 1915, chlorine gas was used for the purpose, but much more dangerous gases were quickly developed. The most effective gas used during World War I was *mustard gas* (actually an easily boiling liquid, which was given its name by British soldiers from its smell). Mustard gas is a *vesicant* (from a Latin word meaning "blister"); that is, it irritates the skin, reddening it and eventually producing blisters and considerable pain. It is poisonous to breathe and damages the eyes.

An even more damaging gas was synthesized during the war by an American chemist, W. Lee Lewis, and was named *Lewisite*. It wasn't produced in quantity before the end of the war and was never used in combat.

In World War II, poison gas was not used. During that war, however, German chemists, searching for compounds to serve as insecticides, came across certain organic phosphates that would kill insects.

Indeed, they were highly effective against any form of life because they interfered with an enzyme called cholinesterase. They prevented cholinesterase from breaking down the compound acetylcholine. Nerve impulses passed from one nerve cell to another by way of acetylcholine, which had to be formed for one impulse and then broken down to be ready to be formed for the next. If cholinesterase stopped working, and acetylcholine didn't break down, various nerve impulses stopped, including those to the heart and lungs. Inhaling small quantities of these odorless organic phosphates meant death within five minutes. Because they interfere with nerve action, they are called *nerve gases*. They are by far the most terrifying poison gases yet developed, but they have never yet been used in actual warfare.

Neutrino

THE FIRST TWO SUBATOMIC PARTICLES discovered, the electron and the proton, carried electric charges. In 1932, a subatomic particle was discovered that carried no charge. It was electrically *neutral* (from Latin words meaning "neither one nor the other"). Physicists made use of the "-on" suffix of the electron and proton and named the new particle *neutron*.

Meanwhile, in 1931, the Austrian-born physicist Wolfgang Pauli had suggested the existence of a new particle to account for the fact that there was some energy missing when a radioactive atomic nucleus broke down and emitted an electron. Pauli showed that such a particle would have to have no charge and very little mass.

What could this particle be called? The term *neutron* was already used by the time other physicists began to take Pauli's particle seriously. Pauli's particle had far less mass than the neutron, so the Italian physicist Enrico Fermi suggested it be called a *neutrino*, which in Italian means "little neutron." This name was accepted.

For many years, the neutrino was thought to be a very doubtful particle indeed — very useful in physical theory, but able to slip through matter so speedily and subtly that it could not be detected. If it could not be detected, how could anyone ever tell if it existed?

In the 1950s, however, American physicists, Clyde L. Cowan, Jr., and Frederick Reines, set up an experiment in which hordes of neutrinos (assuming they existed) would bombard large tanks of water. It was calculated that a very few neutrinos would be absorbed and produce effects that could be observed. The predicted effects were indeed produced and, in 1956, Cowan and Reines were able to announce that the neutrino had been detected and that it really existed. Since the 1960s, physicists have been trying to detect neutrinos produced by the sun.

Neutron Activation Analysis

As EARLY AS 1906, the New Zealand–born physicist Ernest Rutherford was bombarding matter with subatomic particles. If such subatomic particles struck the tiny nucleus of an atom, they could bring about changes in the nuclear structure and change one type of atom into another.

The first subatomic particles that were available for bombardment carried a positive electric charge and were repelled by atomic nuclei which were also positively charged. As a result, fewer subatomic particles managed to strike a nucleus than one might hope.

The situation was changed when the neutron was discovered by the English physicist Sir James Chadwick in 1930. The neutron carried no electric charge and it was not repelled by any part of the atom, either by the positively charged nucleus or by the negatively charged electrons that surrounded it. Quickly, the neutron became a favorite bombarding particle and almost every type of atom was exposed to its action.

Sometimes, a particular stable nucleus would absorb a neutron and become a more massive but still stable nucleus, as when carbon 12 became carbon 13. Much more often, a stable nucleus would absorb a neutron and become radioactive. The new radioactive nucleus would break down to emit certain particles at certain energies. Every different radioactive nucleus would break down in its own special way, different from that of all others. The speeding particles it gives off can be detected with great accuracy. In this way, one can detect what nuclei were present to begin with. This method of analyzing is called *neutron activation analysis*.

Neutron activation analysis is very delicate; amounts as small as a trillionth of a gram of a particular type of nucleus can be detected. Tiny scraps of paint from a work of art can be studied to see if it is a fake. Even hair from Napoleon's century-and-a-half-old corpse was studied and found to contain suspicious quantities of arsenic.

157

Neutron Star

In 1968, pulsars were discovered, objects that gave out very short, regular bursts of radio waves at second-or-less intervals. Naturally, the question was: what could produce such bursts? Some astronomical body must be undergoing some change at intervals rapid enough to produce the pulses. Something must be pulsating, rotating, or revolving about another object, and at every pulsation, rotation, or revolution send out a beam of radio waves.

But the regular change must take place so quickly that the only way to account for it was to suppose some very small body was involved and that it was affected by a very intense gravitational field. The smallest, most intensely gravitational bodies known were white dwarfs, stars in which ordinary atoms had broken down into a mixture of electrons and nuclei. Without their electron shields, the stars collapsed, the nuclei approaching each other far more closely than they would in ordinary matter.

A white dwarf would have the mass of the sun compressed into a ball not much larger across than Earth. Even so, astronomers couldn't see any way of having a white dwarf pulsate, rotate, or revolve quickly enough to account for the pulsars.

Could there be still smaller objects? Suppose the shattered atoms of a white dwarf collapsed further. The negatively charged electrons present would be forced by enormous pressures to combine with the positively charged protons to form uncharged neutrons. The neutrons would smash together to form solid *neutronium*. A star like our sun could condense into a small sphere, ten miles across, retaining all its mass. Such a tiny, but extremely dense *neutron star* could rotate about its axis rapidly enough to account for the pulsars. The theory that pulsars are rotating neutron stars is, in fact, generally accepted now.

Nimbus

THE FIRST *weather satellites* belonged to the TIROS series. They worked most successfully, but they did have certain limitations. Their cameras did not point to Earth all the time, so that pictures of Earth were taken only at intervals. Furthermore, the orientation of the TIROS satellites wobbled somewhat, so that it shifted from Northern Hemisphere to Southern and back again. Then, too, their orbits were such that they were always to be found over the tropical or temperate zones and the polar areas were never properly viewed.

On August 28, 1964, the first of a new type of weather satellite was launched. It was sent into an orbit with a high inclination to the equator, so that it could take photographs of the polar regions, too (these were very important in any consideration of global weather patterns). Furthermore, it was oriented in such a way that its television camera always pointed to Earth as it moved, so that pictures could be taken continually.

The new satellite could also sense infrared radiation given off by Earth. This radiation is given off by the night side and its amount varies according to whether a region of the surface is covered by snow, by ice, by clouds, or by none of these. The infrared data can therefore be used to study the night side of the planet as well as the day side.

Because of the efficiency with which the new satellite studied the cloud cover, it was called *Nimbus I* from the Latin word for "cloud."

By 1966, there were satellites so designed that a picture of the entire planet Earth and its cloud cover could be taken bit by bit, every twenty-four hours. The pictures can be put together to form a large mosaic, and weather stations now use satellite data routinely, as a further refinement in weather prediction.

159

Noble Gas Compounds

In 1894, a Scottish chemist, Sir William Ramsay, and an English physicist, Lord Rayleigh, discovered a gaseous element which did not react with any other substance to form compounds. Its atoms would not join those of any other element to form molecules and it was therefore inert. It was named *argon* from a Greek word meaning "inert."

In the course of the next four years, Ramsay, working with liquid air, located four more gases very like argon in properties: helium, neon, krypton, and xenon. In 1900, still another gas, a radioactive one, was discovered as a breakdown product of radium by a German physicist, Friedrich E. Dorn. He called it *radon*.

The six new elements were lumped together as the *inert gases*. They were also sometimes called the *noble gases*, the point of the adjective being that these gases were too haughty and aristocratic to mingle with the common herd of elements and form compounds with them.

In 1933, however, the American chemist Linus Pauling worked out detailed theories of the manner in which atoms behaved. The inert gases with the larger atoms, he maintained, were not entirely inert and the large atom of xenon ought to combine with the most active of all substances, fluorine.

Xenon was quite rare and fluorine was dangerous to work with, so the suggestion did not come to anything for a while. During World War II, however, techniques for handling fluorine were improved and in 1962, the Canadian chemist Neil Bartlett was able to form a compound with a molecule that included both xenon and fluorine atoms. Other such compounds followed with radon and krypton as well, and the term *inert gases* was dropped. *Noble gases* was used instead (although that was just as much a misnomer). Beginning in 1962, a new and rather unexpected branch of chemistry was founded, that of the *noble gas compounds*.

Nucleon

In 1911, the New Zealand–born physicist Ernest Rutherford demonstrated that almost all the mass of the atom was concentrated in a tiny volume at its center. Outside that volume were one or more very light electrons, nothing more. The tiny massive volume at the center of the atom was called the *atomic nucleus*, from a Latin word meaning "little nut."

By 1914, the existence of the proton was clearly established. It was 1836 times as massive as the electron and every atomic nucleus contained one or more protons. All protons carry a positive electric charge, however, and repel each other. How could the protons all exist within the volume of the tiny nucleus then? At first, physicists decided that some of the negatively charged electrons were also present in the nucleus and by their attraction for protons helped keep the nucleus together.

There were certain theoretical shortcomings to an atomic nucleus made up of protons and electrons only, however. In 1932, Sir James Chadwick discovered the neutron, and it was at once clear that the nucleus was composed of protons and neutrons. The neutrons possessed no electric charge at all, but what held the protons and neutrons together in the nucleus turned out to involve a new force field altogether, the strong nuclear interaction.

Except for electric charge, protons and neutrons are very similar. They have almost identical mass and share many other properties as well. Some term was needed to include both. In 1941, the Danish physicist Christian Møller, noting that both particles occurred in the atomic nucleus, used the "-on" suffix present in the names of subatomic particles and proposed the term *nucleon*. It was quickly accepted. The word *nucleonics* then came to mean the study of the atomic nucleus and its particles generally.

161

Oceanography

As EARLY AS B.C. 3000, the island of Crete had developed a civilization based on sea trade. It was followed by the Phoenicians who, before the time of Julius Caesar, had already explored the Atlantic Ocean as far north as the British Isles and far enough south to have circumnavigated Africa.

Until modern times, however, the ocean was only a highway — only something to cross, and something (in case of storms) to be feared. Men knew virtually nothing about the ocean, except its size. They knew next to nothing about its currents and nothing at all about its depths.

In 1769, the American scholar Benjamin Franklin had taken note of the warm current that moved steadily northeastward along the eastern coast of central North America. Since it seemed to come from the Gulf of Mexico, it was called the *Gulf Stream*.

In the 1850s, a retired American naval officer, Matthew F. Maury, subjected the Gulf Stream and other ocean currents to close study. He gathered data on ocean depths and other material of the sort and, in 1855, published a book on the subject. It was the first major book on what is now known as *oceanography* (short for "ocean geography").

Also in the 1850s, the American businessman Cyrus W. Field was trying to lay a cable across the Atlantic to allow telegraphic communication between America and Europe. This led to interest in the ocean deeps and to the discovery of life there. In 1860, a telegraph cable was brought up from the bottom of the Mediterranean, a mile deep, with the clearest evidence of deep-sea life, for it was found to be encrusted with corals.

Although a century old, it is only now that oceanography is becoming a major science. The study of the ocean bottom has helped us understand the past history of Earth's crust and it is to the oceans, too, that man may have to look for much of his food supply in the future.

Omega-Minus

By 1960, the number of different subatomic particles that were known had passed the hundred mark and was still climbing. Physicists could not help but wonder why so many particles existed. Could the particles exist in families? Perhaps within each family particles existed in different varieties but were essentially the same in many respects. It would be easier to account for a small number of particle families than for a large number of individual particles.

In 1961, the American physicist Murray Gell-Mann and the Israeli physicist Yuval Ne'eman each suggested a way of building such families by making use of eight different properties of the various particles.

For instance, Gell-Mann prepared a kind of triangle of particles: four *delta particles* at the bottom, three *sigma particles* above them, two *xi particles* above them, and at the apex a single particle. The properties all varied in a regular pattern. The four delta particles had electric charges of −1, 0, +1, and +2, in that order. The three sigma particles had charges of −1, 0, and +1, and the two xi particles, charges of −1 and 0. The topmost particle would naturally have a charge of −1.

The catch in this triangle of particles was that the topmost particle was unknown. All the properties needed to make it fit the triangle could be worked out, though, and in particular it would have to have something called a *strangeness number* of −3. Physicists were skeptical of this, for no particle with such a high strangeness number existed anywhere.

Since the undiscovered particle was the last one in the triangle, Gell-Mann called it an *omega particle*, since *omega* was the last letter in the Greek alphabet. Because it should have a negative electric charge, he called it *omega-minus*. Physicists began an intensive search for it in bubble-chamber tracks. Since they knew all the properties it was supposed to have, they knew what to look for and, in 1964, they found it!

Orbital

In 1906, the New Zealand–born physicist Ernest Rutherford showed that atoms had a positively charged core, the *atomic nucleus*, with light electrons in its outskirts. The atoms seemed to be a very tiny solar system, with electron planets circling in orbits about a nucleus sun.

By nineteenth century views though, electrons, moving rapidly in a curved path about the nucleus, should radiate energy away and spiral into the nucleus — yet they did not.

In 1913, the Danish physicist Niels H. D. Bohr suggested that electrons did *not* radiate energy while in orbit, but only when they changed orbits. When they moved into a larger orbit farther from the nucleus, they absorbed a fixed quantity of energy; when they dropped to a smaller orbit, they emitted that energy. This fitted the new *quantum theory* advanced in 1900 by the German physicist Max Planck.

Further investigations into atomic structure showed that electrons could not be viewed as merely particles. In 1926, the Austrian physicist Erwin Schrödinger pictured subatomic particles as waves. The electron was a wave form that spread all about the nucleus like a fuzzy shell. Under certain conditions, the wave form was concentrated in one section or another of the region outside the nucleus, and in 1932, the American chemist Linus Pauling showed how this sort of *wave mechanics* could be used to account for the atom's chemical behavior.

With the electron as a wave form, concentrated here and there, one could not consider it as moving in an orbit. One had to think of the various electrons in an atom as possessing certain energy levels instead. The old word *orbit* was not completely discarded, however, merely changed. Now physicists and chemists speak of electron *orbitals* to distinguish the different energy levels.

164

Organelle

WHEN CELLS were discovered in the 1830s, they seemed to contain undifferentiated protoplasm. Microscopes were too poor to see much detail within the nearly transparent cell.

Microscopes improved, however, and biologists developed the technique of adding colored chemicals which attached themselves to some parts of the cellular interior and not others — making parts of the interior clearly visible. The cell had already been found to possess a central *nucleus* (from Latin words meaning "little nut"). The part of the protoplasm inside the nucleus is the *karyoplasm*, the prefix coming from the Greek word for "nut." The part outside is the *cytoplasm*, the prefix coming from the Greek word for "cell."

Both the karyoplasm and the cytoplasm were found to contain smaller objects of definite shape. The karyoplasm contained the chromosomes, for instance, while the cytoplasm contained mitochondria, ribosomes, and so on. These smaller objects had definite functions within the cell, so it seemed that just as the functions of an organism were divided among various organs, those of a cell were divided among these particles. The particles therefore came to be called *organelles*, from a Latin word meaning "little organ."

The larger and more complex organelles, such as the mitochondria, have been found to possess small quantities of DNA, a substance characteristic of the nucleus. This has caused some biologists to wonder if perhaps the modern cell is really a complex combination of simpler, once independent objects. Perhaps in the beginnings of the development of life, there were living subcellular fragments, whose chemistry was less complex than that of modern cells. Individual organelles may have combined to form a cell more efficient than they themselves. What we see now as organelles may represent, therefore, the remains of precellular life forms.

165

Pacemaker

THE HUMAN HEART beats steadily and rhythmically, hastening or slowing its beat when necessary, but by and large going about its work with calm perseverance for as long as a hundred years or more.

This heart rhythm is a built-in property of heart muscle, and a heart, taken out of the body, isolated from all nerve stimulation, but immersed in a solution containing the proper atoms and atom groups in the proper proportions, will continue beating if that solution is pumped through the heart's blood vessels.

Not only will intact hearts beat in this fashion but even portions of hearts will. It was found, in fact, that different parts of the heart will beat at different rates if they are studied in isolation. In the intact heart, however, the part that beats most rapidly forces its rate on all the rest. The most rapidly beating part of the heart is therefore referred to as the *pacemaker*.

In the human being, the pacemaker is located in a bundle of special cells in the right auricle. On occasion, the pacemaker in the human heart ceases to function properly. This does not mean the heart stops beating altogether; it merely starts beating at the rate of the next most rapid natural rhythm. Naturally, the heartbeat is slowed and a person so afflicted is less able to indulge in hard work or hard exercise.

With the coming of miniaturized electronic devices after World War II, it became possible to manufacture small objects designed to yield a periodic electric discharge that would stimulate the heart to beat at something approaching its normal rhythm. This could then be implanted in the heart surgically as an *artificial pacemaker*. By the end of the 1960s, many thousands of human beings were living normal lives, thanks to such devices.

166

Pair Production

According to the special theory of relativity, evolved by Albert Einstein in 1905, mass and energy are different aspects of the same thing and one can be converted into another. Mass is an extremely concentrated form of energy, and a tiny bit of mass can be converted into a large amount of energy, as in a nuclear bomb. The reverse effect, the conversion of energy into matter, requires that a great deal of energy must be concentrated into a tiny volume. Even then, only a small fragment of matter is formed.

What's more, matter can only be produced if certain of its properties are conserved. For instance, an electric charge cannot be created out of nothing. To create an electron, with its negative electric charge, out of energy, a second particle with an equal positive electric charge must also be formed. The negative and positive charge, taken together, add up to zero and therefore do not count as the creation of electric charge.

It is possible, then, to take an energetic gamma ray photon, and, under the proper conditions, convert it into an electron-positron pair. It is this *pair production*, first observed in 1933 by the British physicist Patrick M. S. B. Blackett, that is the clearest example of the conversion of energy into matter in practice.

Under the influence of a magnetic field, the positron curves off in one direction and the electron in the other. As soon as a positron meets an electron (and our universe is full of them), it combines with it, and the matter of the two particles is converted into energy again. This is *mutual annihilation*. Occasionally, the positron and electron glancingly approach and circle each other for something less than a millionth of a second before annihilation takes place. The combination of circling particles is called *positronium*.

167

Paleobiochemistry

FOR CENTURIES NOW, men have been coming across stony objects in rock which resemble organisms. These *fossils* are the petrified bones, teeth, shells — and sometimes traces of the soft tissues, too — of creatures that have been long dead. Some fossils are over half a billion years old.

The study of these fossils has made it possible to work out details of evolution. The evolution of the horse is well worked out, for instance, and the lines of descent of the ancient reptiles as well.

Fossils tell us only about the superficial appearance of past species. They don't ordinarily tell us much about the more intimate details: the colors of a dinosaur's scales; or whether a particular reptile might have been warm blooded or had begun to develop hair.

It has become possible, however, to obtain remnants of organic compounds from some ancient bones. Amino acids have been extracted and were found to be the same amino acids that occur in proteins today. That at least has not changed.

In the 1960s, scientists went further in this field of *paleobiochemistry* (from Greek words meaning "the chemistry of ancient life"). From rock formations of the type that would possess fossils if they were younger, the American botanist Elso S. Barghoorn extracted organic compounds that might be remnants of very ancient life.

In rocks two to three billion years old, for instance, hydrocarbons have been extracted that might have originated through the breakdown of chlorophyll. Other hydrocarbons have been found that seem related to certain colored compounds, called terpenes, that are characteristic of plant life.

Simple life forms, not possessing the kind of complex structures that will easily fossilize, thus make their presence known much more subtly, and life on this planet is now thought to be at least three billion years old.

Paleomagnetism

EVER SINCE about 1600, scientists have been measuring changes in Earth's magnetic field. Not only do the magnetic poles shift position with time, but the intensity of the field varies, too. One might suppose, though, that there was no hope of determining the changes that took place prior to 1600.

Fortunately, certain minerals, when they crystallize out of the melted state, line up their crystals in the direction of the magnetic field and in a manner governed by the intensity of that field. Once the crystallization is complete, later changes in direction and intensity have no effect. It is possible, then, to measure the age of certain rocks through the radioactive breakdown of uranium, for instance, then study its crystalline formation carefully, and discover the direction and intensity of Earth's magnetic field at some specific time in the long-distant past.

This new science of *paleomagnetism* (the prefix is from a Greek word meaning "ancient") has developed mightily after the mid-twentieth century. It was discovered, for instance, that every million years or so, there is a long-term change in polarity. The north magnetic pole becomes a south magnetic pole and vice versa. This is not because the poles move bodily across the surface of Earth from one polar region to the other. Apparently, the magnetic field gradually decreases in intensity to zero and then begins increasing again in reverse.

During the period of zero field, cosmic ray particles and other energetic, charged radiation from space are not trapped by the field and gradually decanted into the polar regions. Instead, they strike Earth's surface generally. What effect this would have on life forms is uncertain. Some think it may explain those periods in Earth's history when many species seem to die out over a comparatively brief period of time.

169

Parapsychology

HUMAN BEINGS gain their awareness of the universe about them through certain senses, the chief ones being sight, hearing, smell, taste, and touch. There are other senses, too. The nerve endings in the skin do not merely react to touch; some react, particularly, to hot, cold, or pain. There are also senses within ourselves that let us know the amount of tension on particular muscles so that we know where every part of our body is, at a given moment, or how it is oriented to the pull of gravity.

Do we know, in detail, all the different ways in which we might gain awareness of the universe? Are there undiscovered senses? Can we somehow detect a magnetic field, or radio waves, or cosmic ray particles, or properties of the universe we haven't yet studied and don't know exist?

There are occasionally reports of people who become aware of something that is happening at a distance; or of something that has not yet happened. The former is *clairvoyance* (a French word meaning "clearsightedness"), the latter, *precognition* (from Latin words meaning "to know before"). There are also cases where one person seems to know what another is thinking. This is *telepathy* (from Greek words meaning "to feel at a distance").

All such matters are lumped together as *extrasensory perception*, commonly abbreviated *ESP*. The Latin prefix "extra-" means "outside of," so that ESP is anything perceived outside the senses — at least outside the senses as we know them. Presumably, ESP is the result of senses we haven't properly analyzed.

ESP came into prominence when the American psychologist Joseph B. Rhine began a scientific study of such phenomena. He wrote a book called *Extra-sensory Perception* in 1934. The scientific study of ESP is called *parapsychology*, where "para-" is a Greek prefix meaning "beyond."

Parity

Suppose we give each subatomic particle a label of either A or B. Suppose that when an A particle broke up to form two particles, the products were always either both A or both B. We could then write $A = A + A$ or $A = B + B$. And if a B particle broke up to form two particles, suppose one were always A and the other B. Then $B = A + B$.

You might also find that if two particles collided and formed three, this would lead to cases of the following kind: $A + A = A + B + B$ or $A + B = B + B + B$, but never $A + B = A + A$ or $A + A + B = A + B + B$.

What does all this mean? Suppose you think of A as any even integer such as 2, 4, or 6; and B as any odd integer such as 3, 5, or 7. Two odd integers always add up to an even ($A = B + B$) and so do two even integers ($A = A + A$). An odd and an even integer always add together to form an odd one ($B = A + B$).

Parity is from a Latin word meaning "equal." Originally, even numbers were said to have parity because they could be divided into two equal numbers ($14 = 7 + 7$). Later, the word *parity* was applied to both even and odd. If two numbers were both even or both odd, they were of the *same parity;* if one was even and one odd, they were of *opposite parity*. Subatomic particles seem to act according to the rules that govern even and odd numbers, and they were said to have parity, also.

The subatomic particles did not change parity in the course of interactions. If the parity of two or more particles added up to the equivalent of even (or odd) before interacting, they added up to even (or odd) after interacting. This is called *conservation of parity*.

In 1956, however, two Chinese-American physicists, Tsung Dao Lee and Chen Ning Yang, showed that the even-odd arithmetic didn't hold for certain types of subatomic interaction governed by the weak nuclear field. For these, the law of conservation of parity was broken.

Perfect Number

THE ANCIENT GREEKS enjoyed playing games with numbers, and one of them was to add up the factors of particular integers. For instance, the facts of 12 (not counting the number itself) are 1, 2, 3, 4, and 6. Each of these numbers, but no others, will go evenly into 12. The sum of these factors is 16, which is greater than the number 12 itself, so that 12 was called an *abundant number*. The factors of 10, on the other hand, are 1, 2, and 5, which add up to 8. This is less than the number itself, so 10 is a *deficient number*.

But consider 6. Its factors are 1, 2, and 3, and this adds up to the number itself. The Greeks considered 6, therefore, a *perfect number*. Throughout ancient and medieval times, only four different perfect numbers were known. The second was 28, the factors of which are 1, 2, 4, 7, and 14. The third and fourth perfect numbers are 496 and 8128. The fifth perfect number was not discovered till 1460 and the name of the discoverer is not known. It is 33,550,336.

There are no practical uses for the perfect numbers; they are merely a mathematical curiosity. Mathematicians, however, are curious and have worked out formulas that will yield perfect numbers if certain conditions are met. If even the fifth perfect number is over thirty million, those still higher are, you can well imagine, terribly tedious to work out.

The break came with the development of computers during and after World War II. To demonstrate what computers could do, one could set them to work solving formulas for perfect numbers. By now, twenty-one perfect numbers are known. The twenty-first perfect number, worked out in 1971, is a number with twelve thousand and three digits.

Such a number has no more practical use than any smaller perfect number, but wouldn't the Greeks have been astonished if they could have seen it!

Pesticide

ANIMALS COMPETE with each other for food, and man is included in this contest. Man gained an advantage when he invented agriculture. He grew the plants he desired, such as various grains, in greater quantities and more thickly than they occurred in nature. Other plants which grew in the area he found undesirable and uprooted.

But the existence of solid banks of a particular plant served also to increase the numbers of those nonhuman creatures who fed on it. In a state of nature, insects which fed on the occasional clumps of growing grain would be moderate in number. With acre after acre of nothing but grain, such insects would have an enormous food supply for the taking and they would multiply tremendously. Farmers had to fight hordes of insects that would not have existed in such numbers before the days of agriculture. They would also have to fight fungus infections, or small animals such as rabbits, moles, and crows who would want to feed on the plants he was growing. When he began to herd animals, he had to fight off foxes, weasels, wolves, and other predators.

All the living creatures — plant, animal, and microscopic — that competed with man were dangerous to his food supply and sometimes to his own health. Rats not only preyed on his stores of grain but carried lice which transmitted typhus, bubonic plague, and other diseases.

In recent times, man has developed chemicals that kill the competing creatures while leaving desirable organisms comparatively unharmed. There are *herbicides*, for instance, from Latin words meaning "plantkillers," to remove weeds. There are also *insecticides*, *fungicides*, and so on. Undesirable organisms can be lumped together as *pests*, from a Latin word meaning "plagues." In consequence, chemicals that kill undesirable creatures of any sort have been lumped together now as *pesticides*.

173

Pheromone

HUMAN BEINGS can communicate by talking. Through sounds, gestures, written symbols, abstract ideas can be transmitted from one person to another. Human beings are unique in this respect.

Yet other creatures must be able to communicate in some fashion, if only so that there can be cooperation between two individuals of a species in order that they might reproduce. Within a body, the different parts are made to behave in some cooperative fashion by means of chemical messengers called hormones. Is it possible that chemical messages can be carried on, not only within an organism, but from one organism to another?

Such hormonal effects, carried through water or air from one member of a species to another, are called *pheromones*, the prefix coming from a Greek word meaning "to carry." They are hormones carried over a distance.

Insect pheromones are the most dramatic. A female moth can liberate a compound which will act as a powerful sexual attractant on a male moth of the same species a mile away. Each species must have its own pheromone, for there is no point in affecting a male of another species. Each species must have receiving devices of tremendous delicacy because they must be able to react to just a few molecules in the air.

Pheromones are also used in interspecies conflict. Certain ants raid the nests of other ant species, kidnaping young, which they rear as slaves. The raiders use trails of pheromones which not only aid them in keeping together and coordinating their attacks, but also act to alarm and scatter the ant species they are attacking.

Biologists are laboring to use insect pheromones to lure members of troublesome species to destruction. In this way, they can be absolutely specific, doing no direct harm to any other species.

Phosphor

SUBSTANCES WHICH ABSORB energy in some nonlight form and then give it off as visible light are said to be *luminescent*, from Latin words meaning "to become light." Sometimes there is a delay in giving off the light and emission continues even after the energy source is shut off. This delayed luminescence is called *phosphorescence*, which also means "to become light," the prefix in this case coming from the Greek word for "light." Luminescence without delay is called *fluorescence* because it was first noticed in connection with a mineral called fluorite.

Substances which display phosphorescence are called *phosphors*. Examples are calcium tungstate and zinc sulfide, the phosphor qualities of which depend on their method of preparation and also on the presence of certain impurities.

Fluorescence came into important commercial use through the researches of the French chemist Georges Claude, beginning in 1910. He showed that electric discharges through certain gases under low pressure could cause them to fluoresce. By putting them into glass tubes that could be twisted into various shapes, he developed *neon light* (neon was one of the gases used) in various colors.

Mercury vapor would also fluoresce under such conditions, producing radiation rich in ultraviolet light. In 1935, after methods for producing phosphors in quantity were developed, tubes coated on the inner surface were used. The ultraviolet light from the fluorescing mercury vapor excited a phosphorescent glow in the phosphor coating with the result that a steady white light was given off. Such *fluorescent lights* consumed less electric energy for a given amount of light, and emitted light of greater whiteness and softness, accompanied by far less heat, than ordinary incandescent bulbs could.

Photosynthesis

ALL ANIMALS eat food. They take large food molecules apart into the small units that compose them, absorb those small units, and put them together again into the large molecules that compose their own tissues. In doing so, however, ninety per cent of the original food is used to produce energy for life processes or else is wasted outright.

If animals ate only each other, there would be constant attrition and soon all would be dead. However, while some animals eat other animals, most animals eat plants. The wastage of animals eating animals is made up for by the fact that the animals that are eaten have usually built their own tissues at the expense of plants. Of course, the eating of plants is wasteful, too. Most of the plant tissue does not end up as animal tissue but is consumed for energy or ends as excreta.

How, then, do plants keep their own tissue volume up to the mark despite the day-in-day-out ravages of hundreds of thousands of species of animals?

Plants use as their raw materials water and carbon dioxide, which are among the excretion products of animals and which are universally available. For the energy required to build up the large tissue molecules out of the small molecules of water and carbon dioxide (plus certain minerals), sunlight is used. Green plants contain a compound known as *chlorophyll* (from Greek words meaning "green leaf"), which absorbs sunlight and makes its energy available for certain chemical changes. This process is called *photosynthesis* (from Greek words meaning "putting together by light").

Though the basic existence of photosynthesis has been known for two hundred years, it was only in the 1950s that the American biochemist Melvin Calvin, making use of radioactive isotopes, actually began to work out the detailed nature of the chemical changes involved.

Pi

ONE OF THE FIGURES in geometry easiest to construct is the circle. Since the distance from the center to all points on the circle is equal, all you need for construction is a compass.

The Greeks discovered that the curve of the circle is a little over three times as long as the width of the circle. This fact is always true regardless of how large or small the circle is. The Greeks tried to determine the exact ratio of the *perimetron* (meaning "the measurement around") to the *diametron* (meaning "the measurement through"). In English these words become *perimeter* and *diameter*. The best the Greeks could do was the estimate by the mathematician Archimedes in the third century B.C. that the ratio was higher than $3\frac{10}{71}$ but lower than $3\frac{10}{70}$, or roughly about 3.142.

If the diameter is set equal to 1 then the perimeter of the circle is equal to whatever the ratio is (3.142). About 1600, the English mathematician William Oughtred invented the symbol π to stand for that ratio. It is the first Greek letter in the word *perimetron* and is called, in English, *pi*. Ever since, pi has been universally used as the symbol for the ratio of perimeter (or *circumference*, the Latin equivalent, meaning "to carry around") to the diameter.

The value of pi turned up in all kinds of mathematical equations and not only in geometrical formulas. There were numerous attempts to get its exact value. By 1717, an English mathematician, Abraham Sharp, worked it out to seventy-two decimal places, of which most people don't bother remembering more than 3.14159 . . . Actually, no exact value is possible. The decimal goes on forever without ever repeating itself.

In 1955, however, a computer calculated the value of pi to 10,017 decimal places in thirty-three hours. Such a value has no practical use but is of interest to mathematical theorists.

Piezoelectricity

In 1880, two French physicists, the brothers Pierre and Paul-Jacques Curie, discovered that when certain crystals were compressed, an electric potential was developed across them. The harder the pressure, the greater the potential. The Curies called this *piezoelectricity*. The prefix was from the Greek word meaning "to press."

It was found to work the other way, too. If the crystals were placed under an electric potential, they pressed together slightly and became constricted. This is called *electrostriction*.

Later on, Pierre Curie married a Polish chemist, Marie Sklodowska, who became known as the famous Madame Curie. She made use of her husband's discovery in early studies of radioactive substances. Under certain conditions, uranium, in breaking down, could produce ions that would cause a tiny electric flow. Madame Curie measured this flow by balancing it against an electric flow in the opposite direction produced by pressing a crystal. The amount of pressure required for balance gave the measure of the electric flow.

If a crystal is put under the influence of a very rapidly oscillating alternating current, the crystal faces move in and out just as rapidly and produce sound waves. If the in-and-out movement is fast enough, sound waves are produced of such high frequency and such short wavelength that they are ultrasonic. On the other hand, sound waves striking the crystal will produce an alternating current.

Piezoelectric crystals can be used, therefore, to connect electricity and sound. They can be used in microphones to pick up sound and convert it into an electric current, which can then be amplified and turned back into sound. The most common crystals used for this purpose are quartz. Others are ammonium dihydrogen phosphate, barium titanate, and Rochelle salt.

Pinch Effect

ONCE THE HYDROGEN BOMB was invented, men began to wonder if its energies could be controlled. The hydrogen bomb works by the fusion of four small hydrogen atoms into a larger helium atom, so what was wanted was *controlled fusion.* In order to have that, the hydrogen had to be kept together and made to undergo fusion at a slow and steady rate. If this could be done, man would have an enormous energy source that would last millions of years.

A mass of hydrogen is kept together and allowed to undergo steady fusion in the sun and in other stars, but what keeps it together in those cases is an enormous gravitational field. Such a field cannot be produced in the laboratory and other containing methods must be used.

An ordinary container won't work, for hydrogen will not undergo fusion until temperatures of more than a hundred million degrees are reached. At this stage, no material body could possibly contain it. Either the material body would melt and the gas escape, or the gas would cool down upon touching the container. In either case, fusion would stop.

At very high temperatures, however, hydrogen atoms break down into a mixture of electrically charged electrons and protons. Charged particles respond to a magnetic field. If such a field surrounded a tube containing the hydrogen, the particles would be pushed inward. They wouldn't touch the walls of the tube, but would be contained in an invisible *magnetic bottle.* Because the gas is pinched inward, the American physicist Lewi Tonks, who worked out the theory in 1937, called it the *pinch effect.*

A cylindrical magnetic field does not produce a stable pinch effect. The contained particles escape within a few millionths of a second. In order to produce the conditions for controlled fusion, physicists have therefore been working with magnetic fields of more complicated shape.

Pineal Gland

A CONICAL REDDISH GRAY BODY attached to the base of the brain is called the *pineal gland* because it somewhat resembles a pine cone in shape. It is quite small, being only a quarter of an inch long in man.

In the early seventeenth century, the pineal gland attracted considerable attention. The French scientist René Descartes was under the impression that the pineal gland was found only in humans and never in other animals. He maintained that this small scrap of tissue was the seat of the human soul. It was soon found, however, that the pineal occurs in all vertebrates and is far more prominent in some of them than in man.

It was recognized, in more modern times, as merely a gland and in the late 1950s, biochemists at the University of Oregon began work with 200,000 pineal glands obtained from slaughtered cattle. They finally isolated a tiny quantity of substance that, on injection, lightened the skin of a tadpole. It seemed that the pineal produced a hormone, which they named *melatonin* from Greek words meaning "intensity of darkness." It is not known whether this hormone is of importance in the human body.

Still more exciting was the discovery in the 1960s that the pineal gland might be involved in the establishment of circadian rhythms, the way in which the activities of living tissue rise and fall with the alternation of day and night. The pineal was not always hidden deep in the head as in most modern-day vertebrates. There was a time when the pineal was raised on a stalk and reached the top of the skull. Then, it had some of the structure of an eye. A primitive reptile on some small islands near New Zealand has such a *pineal eye* even today. It is particularly pronounced when the reptile is young. The pineal eye cannot really see but it can detect the presence or absence of light and could therefore initiate the circadian rhythm. How the pineal works in those forms of life where it is deep in the skull is still uncertain, however.

Pion

In 1935, the Japanese physicist Hideki Yukawa advanced theoretical reasons for supposing that particles of intermediate size (with masses greater than that of an electron but less than that of a proton) ought to exist. In 1936, the American physicist Carl D. Anderson discovered such a particle and it was eventually named a *meson* from the Greek word meaning "intermediate."

For a while, it looked as though Yukawa's theory had been confirmed, but then some troubles arose. Yukawa's theory predicted the meson would be 270 times as massive as an electron, but Anderson's meson was quite a bit less massive than that. Then again, according to Yukawa's theory, the meson would interact with atomic nuclei very rapidly, but Anderson's meson would not interact with nuclei at all.

In 1949, an English physicist, Cecil F. Powell, was studying cosmic ray particle tracks in the upper Andes. These struck atoms in the atmosphere and produced whole showers of particles. Powell wanted to get as high as possible, hoping he would detect some cosmic ray particles before they had slammed into atoms of the atmosphere. These original particles were called *primary radiation*.

In his studies, however, he uncovered tracks of particles of intermediate size. What's more, they weren't Anderson's mesons at all; they were distinctly more massive. Indeed, they were 273 times as massive as electrons, almost exactly what Yukawa had predicted. What's more, they interacted strongly and rapidly with atomic nuclei. Powell's mesons were the particles that Yukawa had predicted.

To distinguish them from Anderson's mesons, the new particle was given the initial *p* (for *primary radiation*, which Powell had been studying). The Greek form of the letter (*pi*) was used and the new particle was called the *pi-meson*, or *pion* for short.

Plasma Physics

ORDINARILY, three states of matter are recognized: solid, liquid, and gas. In solids, atoms or molecules are in virtual contact and maintain their position. In liquids, the atoms or molecules are still virtually in contact, but they slip and slide past each other freely and do not maintain any given position. In gases, the atoms or molecules are virtually free of each other and move independently, with large spaces between. In all three states, the atoms (whether singly or in molecular groupings) are whole and complete.

As the temperature rises, the atoms themselves begin to break down, losing electrons, which carry a negative electric charge, and leaving behind atom residues that carry a positive electric charge.

Once a gas is largely or entirely composed of electrically charged particles, its properties are different in many ways from ordinary gases. It can be manipulated by magnetic fields, for instance, as ordinary gases cannot. It is, in fact, a fourth state of matter.

The American chemist Irving Langmuir chose a name for this fourth state in the early 1930s, and chose a rather poor one. In Greek, the term *plasm* was used for something that had a form, a definite shape. In the nineteenth century, the German anatomist Max J. S. Schultze used the term for the liquid part of blood, which had no shape but within which objects with a shape were to be found. This liquid part has been called *blood plasma* ever since, directly reversing the meaning of the word. Langmuir compounded the error by applying *plasma* to the formless mass of charged particles.

In trying to control hydrogen fusion, physicists must work with gases at extremely high temperatures of a hundred million degrees or more. The plasma that results must be held in place by magnetic fields. *Plasma physics* has thus become an extremely important branch of science, one on which mankind's future energy needs will depend.

182

Pleasure Center

SINCE THE BRAIN is the controlling organ of the body, it is a natural belief that different parts of the brain might control different parts of the body. Some thought that by studying the bumps on the skull, one might determine overdeveloped portions of the brain, and deduce specially prominent character traits. This enticing but worthless thought gave rise to the folly of *phrenology* (from Greek words meaning "study of the mind") in the nineteenth century.

A closer study came in 1870, when two German physiologists, Gustav T. Fritsch and Eduard Hitzig, stimulated different portions of the cerebral cortex of a dog in order to note what muscular activity, if any, resulted. It was also possible to destroy a patch of the cortex and take note of what paralysis might or might not result. In consequence, parts of the brain were indeed associated with particular muscle systems. Other parts of the brain were associated with sensations received from particular nerve endings.

Something a little more subtle had come about in 1861, when a French surgeon, Pierre P. Broca, had noted that a particular area of the brain controlled not only the ability to speak but also the ability to understand speech. Damage there caused *aphasia* (from Greek words meaning "no speech").

In 1954, the American physiologist James Olds discovered something still more astonishing, a specific region in the brain which, on stimulation, apparently gave rise to a strongly pleasurable sensation. An electrode fixed to the *pleasure center* of a rat, which the animal could stimulate itself, was stimulated up to 8000 times an hour for days at a time, to the exclusion of food, sex, and sleep. All the desirable things in life may be desirable only insofar as they stimulate the pleasure center. Direct stimulation would make all else unnecessary.

183

Polarized Light

Light consists of waves which undulate in all directions. When light passes through certain transparent crystals, the regular rank and file of atoms in the crystal make it possible for light to undulate in only two directions, at right angles to each other.

These two wave forms behave differently as they pass through the crystal. Both are bent, or *refracted* (from Latin words meaning "to break back"), but by different amounts. One beam goes into the crystal but two come out, in a phenomenon called *double refraction.*

The English scientist Sir Isaac Newton knew, about 1700, of the phenomenon of double refraction but thought that light consisted of small particles rather than of waves. In an attempt to explain double refraction, he speculated that the particles consisted of two types which were separated in double refraction. If they were, this might be analogous to the north and south poles of magnets.

In 1808, a French physicist, Etienne L. Malus, studied double refraction and called each separate beam, consisting presumably of light of a single pole, *polarized light.* This was a poor name, for Newton's notion was soon shown to be wrong, but it stuck.

Reflection often polarizes light and it would be handy if one could find a substance which would let through light polarized in one direction but not light polarized in the other. That would eliminate much of the reflected glare without too much general darkening. Some organic crystals could do this but they were too fragile to use as spectacles, for instance. In the 1930s, though, a Harvard undergraduate, Edwin H. Land, conceived the idea of getting the organic crystals properly lined up and then embedding them in clear plastic. The plastic, once it hardened, would do the job and be strong enough to be used as spectacles. He called the product by the trade name *Polaroid.*

Pollution

EVERY ORGANISM produces waste products no longer useful to itself and which, indeed, if allowed to accumulate will be harmful. In every case, though, the waste products are of use to other forms of life, which often restore it to a form useful to the original waste producer.

Thus, all animals make use of oxygen in air, or dissolved in water, combining it with the carbon in foodstuffs, and excreting carbon dioxide as a waste product. No animal can live in an atmosphere with too great a carbon dioxide content. Plant life, however, can utilize carbon dioxide and, using the energy of sunlight, build it up to foodstuffs again, excreting oxygen as a waste product. Of course, animals can make use of the oxygen again.

In this way, there is an oxygen–carbon dioxide cycle, and through the activities of both plants and animals, both oxygen and carbon dioxide remain in a constant concentration in the atmosphere. There is also a nitrogen cycle, a water cycle, and so on.

For billions of years, such cycles have remained more or less in balance, but man's coming has made a difference. Human agriculture upset the balance in some ways, but with the coming of man's industrial civilization, the upset became truly dangerous. Wastes are produced in quantities so great that other forms of life can't handle them quickly enough. Some wastes, such as nonrusting metals, plastics, and so on, can't be restored to circulation at all. Some wastes are actively poisonous.

Wastes which cannot be comfortably cycled or which are actively poisonous are called *pollution*, a well-known English word from a Latin one meaning "to render thoroughly filthy." As the 1970s opened, the new application of this old word came to stand for a growing nightmare to mankind.

185

Polywater

In 1965, a Soviet scientist, B. V. Deryagin, studying liquid water present in very thin glass tubes, was astonished to find it had unusual properties. It was 1.4 times as dense as water ought to be; it could be heated to 500° C. before it boiled (instead of 100° C.), and it could be cooled to -40° C. before it froze (instead of 0° C.).

American chemists were dubious at the news, but when they repeated Deryagin's work, they came out with the same results. It didn't seem possible that so common a substance as water could present surprises, but it did.

Explanations were at once advanced. The molecules in ordinary water form weak attachments to each other, with the hydrogen atom of one molecule lined up with the oxygen atom of a neighbor to form a *hydrogen bond*. This represents a weak attachment in which two atoms are farther apart than is usual for atoms held together by chemical bonds.

In the narrow space within a hairlike glass tube, the water molecules seemed to be forced closer together. The hydrogen bond attraction, exerted over a shorter distance, held neighboring molecules together strongly. Indeed, still more water molecules joined and clung till what amounted to a large molecule made up of units of thousands of ordinary water molecules were thought to have been built up.

When a large molecule is made up of numerous identical small ones bound together, the small molecules are said to *polymerize* (from Greek words meaning "many parts"). The new form of water was therefore called *polymerized water* or *polywater*, for short.

Some chemists still doubted. They suspected that water might dissolve some of the chemicals in glass and that it was this solution, not pure water, that had the unusual properties. Evidence accumulated that this was so, and after creating excitement for several years, polywater faded away.

Probe, Space

IN ORDER TO BE LAUNCHED into orbit, a satellite must be sent into space at a velocity of at least five miles per second. At lower velocities, it would fall back to Earth before having circled the planet even once. The greater the speed it attained, the farther away from Earth it would loop as it circled.

At velocities of more than seven miles per second, it would move so rapidly away from Earth that it would outstrip the ability of the planet to pull it back. The Earth's gravitational field grows less intense with distance and the speeding satellite would be subjected to less and less pull. Though its speed would decrease, it would never decrease to zero; therefore it would never return. The satellite would have exceeded the *escape velocity* for Earth.

Such a satellite would not, however, be free of the sun's much greater gravity. Though not circling Earth, it would move out into an orbit about the sun and would become an *artificial planet*. The first such artificial planet was launched on January 2, 1959, by the Soviet Union.

The term *artificial planet* is not used very often. A satellite circling Earth sends back information about every point of its rapid orbit. An artificial planet circling the sun usually can send back information for only part of its mighty orbit, which takes a year or more for each revolution. Usually, such a device is sent toward some other world with hopes that it will send back information about that world or its vicinity. It is a *space probe* (from a Latin word meaning "examination"). It examines the body toward which it is launched.

The first artificial planet was launched toward the moon. It was the first *lunar probe*. Since then, the Soviet Union and the United States have launched space probes toward Venus and Mars.

Propellant

THE WORD *propel* is from Latin words meaning "to drive forward." Something which drives a vehicle forward is a propeller, and the term is used for a mechanical device that turns rapidly and forces water or air backward. The backward motion of water or air forces a ship or plane forward.

It is also possible to drive an object in one direction if fuel is burned and if the gases formed are expelled in the opposite direction. Suppose you have a hollow cylinder with a pointed end (for streamlining) and a stick trailing off behind to give it stability. (This is shaped vaguely like a distaff, a roll of linen or other twine held on a stick, from the days when women spun their own twine. The cylinder is therefore called a *rocket*, from an Italian word for "distaff.")

If the cylinder is filled with gunpowder which is ignited, the gases that form push out the rear opening and the rocket flies with increasing rapidity in the other direction. The gunpowder is therefore a propeller, though in this case, another form of the word, *propellant*, is employed.

Gunpowder was the first rocket propellant used, but in the twentieth century, much more powerful and efficient propellants were developed. If a rocket is to penetrate beyond the atmosphere, it must carry its own oxygen (or the equivalent) with it. In 1926, when Robert H. Goddard fired the first rocket potentially capable of going beyond the atmosphere, he used a mixture of gasoline and liquid oxygen as the propellant. This was a *bipropellant* (two propellants) because two substances were involved. It is possible to use a single substance which will break down into gases on heating. Acetylene is an example. This would be a *monopropellant* (one propellant).

Eventually, propellants may involve nuclear reactions.

Prostaglandin

AROUND THE URETHRA in the human male, between the bladder and the penis, is a small piece of glandular tissue, which was observed by the Greek anatomist Herophilus in the third century B.C. He thought of it as standing before the bladder, which is true if you are considering a person lying on his back and look at him from the direction of his feet. Herophilus called it *prostatai adenoidis,* meaning "the gland that stands before." In English that becomes *prostate gland.*

The fluid that surrounds the sperm cells produced by the male is called *seminal plasma.* The word *seminal* is from the Latin word for "seed," since the sperm cells are considered a kind of male seed. The seminal fluid is produced partly by the *seminal vesicles,* two little pouches behind the bladder. *Vesicle* is from a Latin word meaning "a little hollow." The prostate gland also contributes to the seminal fluid.

In the early 1930s, various scientists noticed that something in seminal plasma stimulated smooth muscles and lowered blood pressure. This was what one would expect of a hormone. Since it was thought that such a hormone would be formed by the prostate gland, the Swedish physiologist Ulf S. Von Euler named it *prostaglandin* in 1935.

In 1960, the hormone was actually isolated, with several tons of sheep vesicular glands used as starting material. Prostaglandin was found to exist in numerous varieties (at least six occur naturally in the human body) and has a structure resembling certain fatty acids. They are present in tiny quantities and are manufactured elsewhere than in the seminal vesicles, for they are found in women, too.

Prostaglandins have a variety of actions in the body and are involved in reproduction, in the regulation of blood pressure, in nerve transmission, and so on. In 1971, it was reported that aspirin may exert its pain-relieving action through its effect on prostaglandins.

189

Proteinoid

In 1936, a Russian biochemist, Alexander I. Oparin, published a book, *The Origin of Life*, in which he speculated on the chemical steps that might have led to the formation of the first living organism.

Eventually, biochemists, following Oparin's lead, began to experiment with mixtures of simple chemicals such as those which might have existed in Earth's primitive atmosphere and ocean, and studied the reactions which took place when energy in the form of electrical discharges or ultraviolet light was added. In 1953, an American biochemist, Stanley L. Miller, found that a mixture of simple chemicals, in the presence of electric sparks, formed certain organic acids plus a few amino acids, the building blocks of proteins.

Others joined in the search, and more and more complicated molecules were built up in this fashion. It seemed quite certain that, little by little, given a whole ocean of chemicals under blazing sunlight and with millions of years to work in, complicated proteins and nucleic acids, like those of living tissue, might be built up.

The American biochemist Sidney W. Fox wondered if the process need necessarily be step by step. Once amino acids were formed, might they not join together all at once to form protein? In 1958, he heated amino acids under conditions that might have existed on a hot, volcanic earth and did find that they formed long chain molecules resembling proteins. He called them *proteinoids* (proteinlike).

Fox dissolved the proteinoids in hot water and let the solution cool. He found that the proteinoids collected in tiny spheres about the size of small bacteria. Fox called these *microspheres*.

The microspheres seemed to have some of the properties of cells. They had a membrane, could swell or shrink, and even produce buds. They were not alive, but perhaps they represented a step toward life.

190

Pulsar

In the 1960s, astronomers noted that radio waves from particular sources changed in intensity. In some cases, the intensity was changing very quickly; almost as though it were a radio-wave twinkle. Special radio telescopes were designed to catch the twinkle.

In 1964, the English astronomer Antony Hewish was using such a telescope. He had hardly begun when he detected very brief and very regular bursts of radio energy from a particular place in the sky. Each burst lasted only $\frac{1}{30}$ of a second and came every $1\frac{1}{3}$ seconds. The intervals were so equally and accurately spaced that the value could be worked out to a hundred-millionth of a second. The period of the twinkle he had discovered was 1.33730109 seconds.

Hewish at once searched for other such sources, and by February 1968, he was able to announce he had located four. Other astronomers began to search avidly and more were quickly discovered. In two more years, nearly forty more radio-wave twinkles in the sky were discovered. At first, the astronomers hadn't the faintest notion as to what caused the twinkle. The radio waves came in pulses, so they called the phenomenon *pulsating stars*. By combining the first four and last three letters of the phrase, this became *pulsars*.

The search was on to see if anything could be seen in the spots where the pulsars were located — that is, if they emitted visible light as well. One of the pulsars was located in the Crab Nebula; it pulsed more quickly than any other which had been discovered and there was reason to think this might mean it was the youngest and therefore might be the brightest. Astronomers zeroed in on it and sure enough, in January 1969, it was discovered that a dim star within the Crab Nebula did flash on and off in precise time with the radio pulses. It was the first *optical pulsar* to be discovered.

191

Quark

In 1961, the American physicist Murray Gell-Mann worked out a system for grouping subatomic particles into families. It simplified the jungle of particles that physicists had discovered after World War II. Gell-Mann, however, looked for still further simplification.

Was it possible to imagine a very few particles still simpler than the protons, neutrons, and other ordinary subatomic particles? Could different combinations of these few sub-subatomic particles make up all the different subatomic particles there were? If that were the case, physicists might reach a more basic understanding of the structure of matter.

Gell-Mann did come up with three possible particles (and three corresponding antiparticles) which were particularly unusual because they had fractional electric charges. All charges known to this time were equal to that on an electron (-1) or on a proton ($+1$), or were exact multiples of those charges. On the other hand, one of Gell-Mann's suggested new particles had a charge of $+\frac{2}{3}$ and the other two, $-\frac{1}{3}$. (The equivalent antiparticles had charges of $-\frac{2}{3}$ and $+\frac{1}{3}$, respectively.) A proton might be built up of two $+\frac{2}{3}$ particles and one $-\frac{1}{3}$ to give $+1$ altogether, while a neutron might consist of one $+\frac{2}{3}$ and two $-\frac{1}{3}$ to end up with 0 charge.

To name these particles, Gell-Mann turned to *Finnegans Wake* by James Joyce, a book that contained a riotous series of complicated puns and verbal distortions. One sentence went "Three quarks for Musther Mark" (a distortion, perhaps, of "Three quarts for Mister Mark").

Since it would take three of the Gell-Mann particles to make up each of the more prominent subnuclear particles, Gell-Mann decided to call them *quarks*. The queer name caught on, something that might have astonished James Joyce had he not died in 1941, nearly a quarter-century before his word had entered the scientific vocabulary.

Quasar

AFTER WORLD WAR II, astronomers searched the sky to find regions which were emitting radio waves. Beginning in 1960, they found radio waves associated with what seemed like certain dim and undistinguished stars. Yet why should these emit strong radio-wave beams, while ordinary stars did not? The spectra were studied, and dark lines were discovered which represented certain absent wavelengths in the light the stars emitted. Usually, much information could be gained from these dark lines, but in this case, their nature was a puzzle.

In 1963, the Dutch-American astronomer Maarten Schmidt wondered if the dark lines were those which would ordinarily be located in the ultraviolet region, but were shifted so far toward the long wavelength as to appear in the visible light region. There was usually such a shift toward long wavelength (red shift) in distant objects. The greater the distance, the greater the shift. No one, however, had ever before observed a shift as great as the one Schmidt now suspected. Other radio-emitting stars were studied and they, too, showed this enormous red shift.

The conclusion was that these radio-emitting stars were extremely distant, over one billion light-years away, farther than any other objects in the universe. Clearly, they could not be ordinary stars, for no individual stars could possibly be seen at that distance. Since their nature was unknown, they were called *quasi-stellar objects*, (*quasi-stellar* being a Latin term for "starlike"). In 1964, the Chinese-American astronomer Hong-Yee Chiu used the first and last groups of letters of quasi-stellar and called them *quasars* for short.

The name stuck. As astronomers now think, a quasar is rather small, much smaller than a galaxy, yet it shines with ten times the brilliance of a galaxy. Its exact nature is still in dispute and some astronomers even deny that the enormous red shift can really be interpreted as representing great distance.

Quasi-Mammal

SHORTLY BEFORE 1800, certain primitive mammals discovered in Australia surprised zoologists. Although they had hair, the hallmark of the mammal, they laid eggs, something no other mammals did.

These egg-laying mammals are called *monotremes* (from Greek words meaning "one opening," since in place of the separate openings for feces and urine that all other mammals have, they have one opening for both wastes, and for eggs, too — a situation characteristic of birds and reptiles). The best-known monotreme is the duck-billed platypus.

Mammals are descended from reptiles. In particular, they are descended from a now extinct group of reptiles called *therapsids* (from Greek words which mean "beast openings," because they possess openings through skull bones similar in types to those in the skull of mammals).

The only remains we have of the therapsids are bones and teeth. We don't know anything about their soft tissues and skin. We don't know whether they might not have developed hair and warm-bloodedness even while they were still technically reptiles. Perhaps the duckbill and the other monotremes are even today closer to the ancestral therapsid line than to other mammals.

An American zoologist, Giles T. MacIntyre, studied the problem in the late 1960s. He considered the trigeminal nerve, which leads from the jaw muscles to the brain. This nerve follows a slightly different path in reptiles than in mammals. In the adult duckbills the nerve follows the mammalian path, or seems to. In young duckbills, the path can be more easily followed and in them MacIntyre saw that the nerve traveled the reptilian path. It seems possible, then, that the monotremes might be the last surviving therapsids rather than being the most primitive mammals. MacIntyre speaks of the monotremes as *quasi-mammals*, where *quasi* is a Latin word meaning "having some resemblance to."

Radiocarbon Dating

IN RECENT DECADES, uranium breakdown to lead has been used to date extremely old rocks. Uranium breaks down at a known, very slow rate and from the relative amount of uranium and lead in certain rocks, it can be shown that particular rocks might be a billion years old or more.

Uranium breaks down so slowly, however, that in a rock only a few thousand years old, the amount of lead formed is too small to be detectable. What was needed for shorter date determinations was a substance that broke down more quickly. In 1939, two American biochemists, Martin D. Kamen and Samuel Ruben, discovered a form of carbon, carbon 14, which broke down at such a rate that half was gone in only 5770 years. In breaking down, it gave off radiations of speeding electrons, so it was called *radiocarbon*.

Although carbon 14 breaks down quickly, cosmic ray particles, smashing into atmospheric atoms, are continually producing new carbon 14 atoms. The carbon 14 content of the atmosphere, while very small, therefore remains steady. As long as an organism is alive, it keeps incorporating carbon 14 into its tissues as fast as those atoms break down so there is a steady small content. Once the organism dies, however, incorporation ceases and the carbon 14 present slowly disappears at a steady rate.

In 1946, the American chemist Willard F. Libby developed methods for determining the carbon 14 content in old materials that had once been part of a living organism. Charred wood from ancient campfires, textiles that had been used to wrap mummies, anything of that sort was used. The carbon 14 content was determined, and from that one could tell how long it had been since the material had ceased to be part of a living organism. This procedure was called *radiocarbon dating* and it had helped archeologists define more clearly certain events in the last thirty thousand years, notably the exact times of the coming and going of the great glaciers.

Red Shift

In 1842, an Austrian physicist, Christian J. Doppler, demonstrated that when sound of a particular pitch was emitted by a source that was approaching the observer, the sound was higher than it would have been if the source were motionless. The approaching source squeezed the sound waves together and made them shorter. On the other hand, if the source were moving away, the sound waves were pulled apart and made longer, so that the pitch became deeper.

This *Doppler effect* could apply to any wave phenomenon, to light, for instance. If a light source were approaching us, the waves it gave off would squeeze together and become shorter. If a spectrum were being studied in which the light was separated into different wavelengths, all the wavelengths would shift toward the short wavelength end of the spectrum. That end was occupied by light seen as violet in color, so that the Doppler effect in an approaching light source was called the *violet shift*. If the light source were receding, the wavelengths would lengthen and shift toward the other end, occupied by red light. That would be the *red shift*.

The violet shift and the red shift were easily seen when the spectrum included dark lines because of the wavelengths that were absorbed. The dark lines shifted, and since their proper position was known, the exact extent of the shift could be easily measured.

In the 1920s, it became clear to the American astronomer Edwin P. Hubble that the distant galaxies were all receding from us. In fact, the more distant the galaxy, the more rapid the recession and, therefore, the greater the red shift. With that, the measurement of the red shift became an important way of learning some fundamental things about the universe as a whole. It was by the detection of an unexpectedly enormous red shift that the puzzling quasars were discovered.

196

REM Sleep

Sleep is a necessity. Lack of sleep will kill more quickly than lack of food will. Yet why do we sleep? It can't be merely to rest, for we might just lie quietly in bed wide awake and that would not substitute for sleep. Indeed, sleep is sometimes full of activity, what with twisting and turning in the bed, so that there is little rest indeed, and yet it will accomplish its purpose.

When wakefulness is enforced, no bodily functions go seriously awry except those of the brain. Extended wakefulness affects the coordination of various parts of the nervous system, and there is the onset of hallucinations and other symptoms of mental disturbance. So sleep must do something for the brain that mere wakeful rest does not.

What about dreams? Few people seem to think of dreams as being of physical importance, and indeed, a dreamless sleep is sometimes spoken of as particularly restful. Yet dreamlessness merely means not remembering dreams, not the lack of them.

The American physiologist W. Dement, studying sleeping subjects in 1952, noticed periods of rapid eye movements that sometimes persisted for minutes. He called this *rapid-eye-movement sleep* or, abbreviated, *REM sleep*. During this period, breathing, heartbeat, and blood pressure rose to waking levels. This occupied a quarter of the sleeping time. If a sleeper was awakened during these periods, he generally reported being in the midst of a dream. Furthermore, if a sleeper was continually disturbed during these periods, he began to suffer psychological distress and periods of REM sleep multiplied during succeeding nights, as though to make up for the lost dreaming.

Apparently, it is not just sleep the brain needs if it is to recover from the wear and tear of the day, but REM sleep. Dreams somehow restore the proper functioning of the nervous system.

Resonance Particles

AFTER THE DISCOVERY of radioactivity in the 1890s, it was found that some types of radioactive atoms only existed very briefly before breaking down. Polonium 212 breaks down in less than a millionth of a second. This could be measured by the energy of the particle it emits (the greater the energy, the faster the breakdown).

Subatomic particles themselves could break down. For instance, the muon breaks down after about two millionths of a second; the pion, after only two hundred-millionths of a second; some of the hyperons, in less than a billionth of a second.

These ultrashort intervals can be measured by noting how far a particle travels in a bubble chamber before breaking down. If a particle is formed, travels three centimeters, then breaks down, and if it travels at nearly the speed of light, it would have taken about one ten-billionth of a second to go that distance.

Actually, even a ten-billionth of a second is a long time on the subatomic scale. In the 1950s, it seemed that pions and protons interacted very readily at some energies and not at others. When something happens at some energies and not at others, it is referred to as a *resonance* in an analogy with what happens in sound when some sound waves but not others cause a receiver to *resonate* (from Latin words meaning "to sound again") in response.

In the 1960s, the American physicist Luis W. Alvarez maintained that actual particles were being formed in these resonance events and that they broke down in a few trillion-trillionths of a second. Even at the speed of light particles could only travel a submicroscopic distance in this time. The existence of these *resonance particles* could only be demonstrated indirectly, but that evidence is quite convincing, and physicists now accept their existence.

Retinene

Plastered up against the internal surface of the rear of the eyeball is a coating about the size and thickness of a postage stamp. It is called the *retina*, which may come from a Latin word meaning "net," but may not. No one knows the real origin of the name.

The cells of the retina are sensitive to light and enable us to see. They come in two varieties, *cones* and *rods* (so called from their shapes). The cones are stimulated by comparatively bright light and react to the different colors. The rods are stimulated by dim light but do not distinguish between colors.

In 1877, a biologist working in Rome, Franz Boll, reported that the frog retina had a rose color that bleached on exposure to light. The German biologist Fritz Kühne extracted the colored compound the following year. It seemed purplish to him and he called it by a German phrase translated into English as *visual purple*. The more formal chemical name *rhodopsin* is from Greek words meaning "rose eye." Rhodopsin occurs in the rods and makes vision in dim light possible.

In the 1940s, the American biochemist George Wald studied rhodopsin and found that it could be broken into two parts, a large protein portion and a smaller colored compound, or pigment. The protein portion he called *opsin*, the pigment, *retinene*, from "retina."

Retinene, it turned out, is closely related in chemical structure to vitamin A, but is less stable. Retinene undergoes delicate changes in structure which make vision in dim light possible and as some of it is destroyed, more is formed out of vitamin A.

It is for this reason that a vitamin A deficiency in the diet affects vision. The supply of retinene fails and the result is *night blindness*, an inability to see in dim light. This, however, is not the only result of vitamin A deficiency.

Ribosome

WHEN THE EXISTENCE of tissue cells was first recognized in the 1830s, little could be discovered concerning their intricate internal structure. With time, however, new techniques, both chemical and physical, were evolved, and these made the tiny structures within the cell visible. Eventually, for instance, small structures called *mitochondria* were located in the cytoplasm and were found to be the powerhouses of the cells, the structures within which energy-releasing reactions took place.

Then in the 1930s, the electron microscope was invented, which magnified much more than ordinary microscopes. The electron microscope was continually improved and by the 1950s, the tiny mitochondria were in turn found to have a complex structure. In addition, still smaller objects were observed. These smaller bodies were about one ten-thousandth the size of mitochondria and little could be made out concerning them. They were called *microsomes*, from Greek words meaning "small bodies."

About the only thing that seemed remarkable in the microsomes was the fact that they contained quantities of ribonucleic acid, or RNA. This was coming to be considered an important substance in connection with the chemistry of inheritance and interest in them increased. In 1953, the Rumanian-American biochemist George E. Palade found tiny particles densely distributed on the microsome membranes. By 1956, he had isolated these tiny particles (each about a thousandth the size of a microsome) and found that they contained nearly all the RNA of the microsomes. These still smaller particles he therefore called *ribosomes*.

In 1960, it was discovered that it was on the ribosomes that specific protein molecules within the cell were formed and that these tiny objects were therefore a key component of the system by which inherited characteristics were passed from cell to cell and from generation to generation.

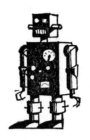

Robot

THROUGH HISTORY, men have dreamed of manufacturing mechanical men who would perform all tasks without ever getting tired or rebellious. In the Greek myths there was a bronze man who protected the shores of Crete and golden girls who helped the smith-god Hephaestus. In the Middle Ages, there were legends of the construction of mechanical men, too.

This view entered the public consciousness most sharply in 1818, when a novel, *Frankenstein*, by Mary Shelley (the wife of the poet Percy B. Shelley) was published. Frankenstein was a student of anatomy who produced a living body out of scraps of corpses. The being so created eventually killed Frankenstein.

In 1921, after World War I with its airplanes and its poison gas had shown how deadly science could be made to be, a more elaborate version of this view was published. The Czech dramatist Karel Capek wrote a play in which a scientist produced large numbers of mechanical men designed to be the world's workers, freeing mankind for other, higher tasks. Unfortunately, the scheme went awry and in the end, the mechanical men wiped out all of mankind.

Capek's scientist was named Rossum and his industrial organization was R.U.R., which was also the name of the play. That stood for "Rossum's Universal Robots," where *robot* was the Czech word for "worker." The play was popular enough to cause the word to enter many languages, including English, to represent a mechanical man designed to simulate human activity and do human work.

Over the last forty years, robots have played a part in many science-fiction stories, though the tendency has been to drop the view that they are inevitable enemies of mankind and to stress the safeguards that will be built into them to make them generally useful. And meanwhile, science gets ever closer to developing at least simple robots.

Schizophrenia

As MEDICAL SCIENCE has succeeded in controlling various physical ailments, mental ailments have come to make up an increasing percentage of those illnesses requiring hospitalization. It is estimated that one American out of ten suffers from some form of mental illness, and of the severe forms the most common is *schizophrenia.*

This name was coined in 1911 by a Swiss psychiatrist, Eugen Bleuler. It comes from Greek words meaning "split mind." Bleuler used it because persons suffering from this disease seemed to be dominated by one set of ideas (or one *complex*) to the exclusion of others, as though the mind's harmonious working had been disrupted and as though that one portion had seized control of the rest.

At least half of all patients in mental hospitals are schizophrenics of one type or another, and it is estimated that one per cent of mankind (perhaps 35 million people altogether) is affected.

Some way of preventing the disease, or curing it, or at least mitigating its effects, would be very desirable. Could it be caused by some vitamin deficiency, for instance? There is a disease, pellagra, brought about by a lack of the vitamin *niacin,* and it produces mental disorders. In the body, niacin forms part of a more complicated molecule called *niacin adenine dinucleotide* (NAD) and there were indeed reports in the middle 1960s that massive treatment with NAD brought about improvement in some schizophrenics.

An older name for the disease was *dementia praecox,* from Latin words meaning "early-ripening madness." This was intended to differentiate it from mental illness affecting the old through the deterioration of the brain with age (senile dementia), since schizophrenia makes itself manifest at a comparatively early age, generally between the years of 18 and 28.

Schmidt Camera

FROM 1609, when the Italian scientist Galileo Galilei first used a telescope to observe the heavens, astronomers have been using better and better instruments for the purpose. Now telescopes with huge mirrors, 200 inches across or more, are in use or are being constructed and with it objects millions of light-years away can be studied.

These huge telescopes have mirrors whose shapes are paraboloids (like that of the rear reflecting surface of an automobile headlight). Only such a surface can exactly focus the light of tiny, distant stars and give a sharp image. However, the larger and more precise a telescope, the smaller the area of the sky that can be focused on.

For the eye this is all right, since the eye can only look at a small area at one time anyway. From the mid nineteenth century on, however, astronomers used photography more and more and it was very tedious to photograph tiny bits of the sky and try to fit all the photographs together. Yet if one were to try to photograph a large section of the sky, it turned out that objects became more and more distorted, the farther the light hit from the center of the paraboloid mirror.

If a spherical mirror was used, the stars wouldn't come to a sharp image anywhere, but it didn't matter where the light hit. The images were equally fuzzy everywhere.

In 1930, an Estonian-born German astronomer, Bernhard Schmidt, designed a correcting lens that could be placed in front of the spherical mirror. This made the stars equally sharp everywhere. With such a *Schmidt telescope*, photographs could be made over large parts of the heavens at once. In fact, such instruments are used so exclusively for photography that they are usually called *Schmidt cameras*.

The largest Schmidt camera now in use has a 53-inch mirror and began operations in East Germany, in 1960.

Scintillation Counter

WHEN SUBATOMIC PARTICLES were first discovered in the 1890s, physicists were at a loss for methods to study them. Since they were so tiny and moved so quickly, it seemed a forlorn hope to be able to detect the effects of a single particle. Yet it proved surprisingly easy to do so. In 1908, the New Zealand–born physicist Ernest Rutherford and his German assistant Hans Geiger allowed subatomic particles to impinge upon a screen of a chemical called zinc sulfide. Every time a particle smashed into the screen, it struck an atom and the energy of its impact was translated into a tiny spark of light, or *scintillation*, from the Latin word *scintilla*, meaning "spark."

By counting the scintillations under a lens, important information could be gathered. This was tedious work, however, and the invention of the cloud chamber, which detected not only single particles but even detected the paths they took, made scintillations obsolete.

In the 1940s, however, both British and German physicists used electronic devices to amplify the scintillations to the point where the amplified light could trigger a counter. In this way, scintillations were automatically counted and the device was naturally called a *scintillation counter*.

Scintillation counters have their special uses. For one thing, they can easily discern types of particles that are detected with difficulty, if at all, by cloud chambers and bubble chambers (which work well only for electrically charged particles). Scintillation counters are very effective in picking up gamma ray photons, for instance, which are uncharged.

Then, too, scintillation counters are very quick in their responses. They can pick up light flashes that last for only a few billionths of a second and are therefore useful in studying events that are very brief.

Scotophobin

ARE MEMORIES represented by special molecules in the brain? What about learning in general? This is an important question. If the brain learns and remembers by reorganizing its connections and setting up special pathways, then what is learned and remembered is peculiar to each particular brain and can never be physically transferred. If special molecules are involved, these might be transferable.

In the early 1960s, experimenters found that a very primitive creature, the flatworm, could be trained to do some simple activity, like responding to light. If these trained flatworms were chopped up and fed to untrained ones, the latter would possess the ability or develop it more easily than otherwise. Some molecule in the trained flatworms seemed to have been incorporated into the untrained, and it meant the same to both.

In 1965, the Danish physiologist Ejnar Fjerdingstad found he could do the same with a much higher animal, the rat. He trained rats to go to light, then injected brain material from these into untrained rats and found that the latter could learn to respond to light more easily than untrained rats which had not been injected.

The Hungarian-American physiologist Georges Ungar went farther. In 1970, he subjected rats to an electric shock in the dark so that they finally developed strong fear of the dark. Brain extracts, when injected into unshocked animals, caused them to show fear, too. From several pounds of brains from animals trained to show fear, he isolated a chemical compound which would induce fear in an untrained animal. This he called *scotophobin* from Greek words meaning "fear of the dark."

The compound is a simple one, made up of a chain of nine amino acids, and it induces fear of the dark not only in rats but in goldfish as well. It seems clear that this simple compound is the closest approach yet to an actual memory molecule.

Sco XR–1

In the 1960s, rockets scanned space for x-ray sources and found the Crab Nebula to be one. The Crab Nebula is the remains of a huge stellar explosion and it is still heaving with energy. Astronomers were anxious to give the source a name according to some methodical system. The Crab Nebula is in the constellation of Taurus and so the x-ray source in it was called *Tau XR–1*, that is, the *first x-ray source to be detected in Taurus*.

The Crab Nebula is by no means the strongest x-ray source in the night sky, however. From the direction of the constellation Scorpio, there comes a beam of x rays eight times more intense than those emitted by the Crab Nebula. This strongest of the *x-ray stars* was named *Sco XR–1*. For some years, despite the probings of astronomers, the sky in the direction of Sco XR–1 seemed blank. But then in 1966, a seemingly ordinary thirteenth-magnitude star was identified as the source. It produced a thousand times as much energy in the form of x rays as of visible light. Why? Nobody knows.

There are other mysteries connected with these x-ray sources. There is a strong source in the constellation of Cygnus, *Cyg XR–1*, which, however, in the course of a single year, lost much of its x-ray intensity. It is a *variable x-ray star*. A second source in that constellation, *Cyg XR–2*, is weaker because it is very far away, perhaps as much as 2000 light-years away. (This was the first x-ray source to have its distance determined — in 1968.) Allowing for its distance, its x-ray emission is a thousand times as energetic as all the radiation, of all kinds, coming from the sun.

A radio source in Centaurus, *Cent XR–2*, came into being in late 1965, got stronger till April 1967, then faded. It was the first *x-ray nova* to be discovered.

Semiconductor

MOST COMMON SUBSTANCES either conduct an electric current very well, like metals, or like glass, porcelain, and sulfur, hardly at all. The former are called *conductors* and the latter, *nonconductors*.

There do exist, however, certain substances, which are rather intermediate in this respect. The elements silicon and germanium conduct a current far more poorly than metals do, but with far more ease than glass does. Such substances are half-conductors, so to speak, or to use the proper Latin prefix, *semiconductors*.

These semiconductors did not rouse great interest until the 1940s, when several physicists — William B. Shockley, John Bardeen, and Walter H. Brattain — began to examine them. They found that germanium and silicon served best as semiconductors when they had small traces of certain impurities. The germanium and silicon had four outer electrons per atom, but a trace of arsenic added an occasional atom with five outer electrons. This fifth electron did not really fit in the orderly array of atoms that made up the silicon or germanium crystal lattice. It drifted from atom to atom under an electric voltage and it was this which carried an electric current from the negative pole to the positive pole. The mobility of the electron increased with temperature so that the conductivity of a semiconductor rose with temperature instead of falling, as was the case with an ordinary conductor.

If a small trace of boron was added to silicon or germanium, something similar happened. The boron had only three outer electrons so that there was a *hole* where the fourth electron ought to have been. A neighboring electron would fill it, forming a new hole which would be filled, forming a new hole again, and so on. Under an electric voltage, this hole traveled from the positive pole to the negative pole, and again the semiconductor effect was produced.

Sequenator

PROTEINS, giant molecules characteristic of life, are made up of long chains of amino acids of about twenty different varieties. In determining the structure of a protein molecule, not only must all the amino acids be identified, but their exact order in the chain must also be determined.

During the 1950s, methods were devised for breaking up the protein chain into short fragments. The particular amino acids making up the short chains were identified and their order worked out. (It is much easier to do this for short chains than for long.) Once this was done, one could reason out the exact structure of the original long chain.

In 1964, the exact structure of trypsin, a protein made up of a chain of 223 amino acids, was deciphered. By 1967, the technique was actually automated. The Swedish-Australian biochemist P. Edman devised a system which could work on as little as five milligrams of pure protein. From the protein chain, amino acids could be peeled off and identified, one by one. In one test, sixty amino acids were peeled off and identified in four days. Edman called the device a *sequenator* because it determined the sequence of the amino acids in the chain.

Once the sequence of amino acids was known, biochemists had begun to try to put amino acids together in the right sequence to form a synthetic version of the protein molecule. This was difficult to do since, each time, the chain had to be dissolved, a particular amino acid added — and the new chain then separated out and dissolved afresh.

In 1959, the American biochemist Robert B. Merrifield used a technique in which an amino acid was bound to beads of a resin. Additional amino acids were added one by one, but at each step, the growing chain was easily separated by filtering out the beads. Synthesizing protein chains became much more efficient. In 1970, the Chinese-American biochemist Choh Hao Li synthesized a 188-amino-acid chain of a growth hormone.

Seyfert Galaxy

ONCE THE VERY DISTANT QUASARS were discovered, astronomers had to determine their nature. To be visible over such great distances, they had to be up to a hundred times as luminous as an ordinary galaxy, yet they were quite small.

This was discovered when, in 1963, they were found to vary in their light emission. The variation was so rapid that the quasars had to be small, since nothing could vary in brightness as a whole unless some sort of impulse could travel from end to end in the time of variation — and nothing could travel faster than light.

It began to seem that a quasar might have the mass of a galaxy, but that the mass was concentrated into a ball only one light-year across or less, instead of the hundred-thousand–light-year stretch across an ordinary galaxy.

It was as though an ordinary galaxy could compress itself into a quasar as an ordinary star could compress itself into a white dwarf. Was a quasar a *white-dwarf galaxy?*

If so, might there be signs of some galaxies that were partway along in the process? Back in 1943, an astronomer, Carl Seyfert, had studied a rather unusual galaxy. Like many other galaxies it had a nucleus and spiral arms, but unlike most others, the nucleus was very bright, as though something very energetic were going on there. Others were found and by 1968, a dozen of these *Seyfert galaxies* were known, though probably they are much more numerous.

Seyfert galaxies are not very distant, but as the process of central compression continues, the center must get smaller and brighter. In the case of advanced Seyfert galaxies that are very distant, the tiny, enormously bright center may be the only thing left visible, and it is that center that we detect (perhaps) as a quasar.

Sickle Cell Anemia

In 1910, an American physician, James H. Herrick, examined a West Indian black with anemia (from Greek words meaning "no blood"), a condition in which the red corpuscles fail to supply adequate oxygen. (The term *anemia* was first used in this way in 1849 by an English physician, Thomas Addison.)

Herrick studied the red blood corpuscles of his patient and found that instead of being normally round and flat, with a shallow depression in the middle, they were distorted into a kind of crescent shape, rather like the blade of a sickle. Herrick called them *sickle cells* and named the condition *sickle cell anemia*. Other people, almost invariably blacks, were also found to suffer in this way and it was recognized as an inherited condition.

In 1949, the American chemist Linus Pauling showed that red corpuscles sickled because they contained an abnormal form of hemoglobin. The abnormal form was named *hemoglobin S*, with the S standing for *sickle cell*, of course.

Hemoglobin S is less soluble than ordinary hemoglobin and it has a tendency to form solid crystals within the red blood corpuscle. It is these crystals that distort the shape of the corpuscle so that it becomes less efficient in transporting oxygen. Sickle cell anemia was the first *molecular disease* to be recognized, the first disease to result from the inheritance of a gene that produced a particular abnormal molecule.

In the 1950s, a German-born biochemist, Vernon M. Ingram, working at Cambridge University in England, analyzed both normal hemoglobin and hemoglobin S and found that each consisted of almost exactly the same chains of over six hundred amino acids. There was a difference in only two amino acids out of all those hundreds, but that was enough to produce a serious disease which few sufferers could survive into adulthood.

Silicone

THE ELEMENT which is most similar to carbon in its electron structure is silicon. Since carbon atoms easily form long chains and complex rings, it might be expected that silicon atoms would do so, too. To a certain extent, indeed, they do, but silicon atoms are larger than carbon atoms and don't form bonds between themselves nearly as strong. This means that chains of silicon atoms tend to be unstable and it is difficult to hook more than a very few of them together.

Silicon atoms, however, form a very tight bond with oxygen atoms, and, indeed, the rocky structure of Earth consists very largely of minerals in which silicon and oxygen atoms are bound tightly together, along with smaller quantities of other atoms, to form *silicates*.

In 1899, the English chemist Frederic S. Kipping began a systematic study of chains of silicon and oxygen atoms in alteration. Such chains could be made of any length and were even more stable than carbon chains. Because carbon and oxygen are bonded together in certain well-known compounds called *ketones*, Kipping used the same ending for his silicon-oxygen chains and called them *silicones*. Each silicon atom in a silicone chain could attach itself to two different atoms or atom groupings and the result was that a vast variety of silicones were possible, depending on the length of the chain and on the nature of the added groupings.

During and after World War II, a variety of silicones were prepared that had useful properties. There were silicone liquids that were used as wetting agents; silicone greases that were used as lubricants; silicone resins that were used as electrical insulators; and soft silicone solids that were used as artificial rubber. In all cases, they had the useful properties of being stable, resistant to heat, and unchanging in properties as temperatures altered.

211

Smog

WE USUALLY THINK of the atmosphere as consisting of gases such as oxygen and nitrogen. Actually, it contains tiny fragments of liquids and solids, too. The liquids are the drops of water formed as clouds, usually high in the air, when water vapor condenses. Sometimes, however, the clouds are at ground level and then we speak of *fog*.

Tiny solid particles enter the air when something burns. This happens even in the absence of any human activity. Lightning, for instance, might set a forest fire that would dump huge amounts of tiny particles of unburnt or partly burnt solids into the air. Volcanic eruptions may send cubic miles of fine, solid material into the stratosphere. After the great Krakatoa explosion of 1883, the dust didn't settle for years. In general, solid particles suspended in air are spoken of as *smoke*.

In the last century, men have burned coal and oil at a steadily increasing rate. If the coal and oil contained nothing but carbon and hydrogen atoms (as they would if they were absolutely pure), then only carbon dioxide and water would be formed if they were well burnt. Combustion usually isn't entirely complete, and carbon monoxide and hydrocarbon fragments are also formed. In addition, sulfates and nitrates are formed from impurities that are present.

The sulfates, particularly, tend to dissolve in any water droplets that might be present in the air, forming an acidic, irritating substance. If the air over a city is stagnant, this combination of smoke and fog (called *smog*) may linger and accumulate, damaging lungs and eyes, making respiratory diseases worse, and sometimes killing those who, through age or illness, cannot tolerate the added stress on their lungs.

The situation has grown much worse since World War II. Killing smogs took place in Donora, Pennsylvania, in 1948, and in London in 1952. Smog is one of the growing problems that science must somehow deal with.

Solar Wind

In 1859 the English astronomer Richard C. Carrington, who was studying sunspots, noted a sudden brightening on the face of the sun. Almost immediately after that observation, disorders were noted in the behavior of the magnetic compass, and the northern lights were particularly brilliant.

Since then, it has been discovered that there are periodic violent eruptions of incandescent matter on the sun, eruptions even hotter than the sun's surface generally. These, which usually take place near sunspots, are called *solar flares*.

The constant heaving of the sun's surface, particularly in flares, sends matter thousands of miles upward and some of it escapes even the sun's giant gravity. As a result, there are quantities of subatomic particles shot out into space. Chiefly because of the flares, the sun is surrounded by matter moving away from it in all directions at speeds of hundreds of miles an hour. About a million tons of matter leave the sun's surface each second, and this steady outward movement of matter has been called the *solar wind*.

It is the particles in this solar wind that are trapped in the magnetic lines of force of Earth, making up the *magnetosphere*. The force of the wind streamlines the magnetosphere, making it spherical and blunt on the side toward the sun and drawn out, like a teardrop, into a long tail on the night side. The solar wind causes the magnetosphere to have a rather sharp boundary called the *magnetopause*.

When a flare happens to shoot upward in Earth's direction, the solar wind toward us is intensified. Unusual numbers of charged particles flood the magnetosphere and enter the upper atmosphere in the polar regions, creating brilliant northern lights, upsetting the compass, and interfering with radio and television reception. This is a *magnetic storm*.

213

Sonar

BATS FIND their way around in the dark by producing high-pitched squeaks that bounce off obstacles in an echo. By listening for the direction of the echo and noting the time lapse between squeak and echo, the bat knows the direction and distance of an obstacle from it, and can thus avoid it.

It seemed reasonable to suppose that men might develop instruments that could take advantage of this principle. What was needed were sounds sufficiently high pitched. The higher the pitch (particularly if it were ultrasonic — that is, too high pitched for the human ear to detect) the further it would penetrate in a particular direction and the more easily it would be reflected.

In 1880, the French physicist Pierre Curie and his brother Paul-Jacques devised a method of producing high-pitched sound waves easily. In the early twentieth century, the system they used was improved to the point where sound waves far in the ultrasonic range could be produced. During World War I, a French physicist, Paul Langevin, applied these ultrasonic sound waves to the detection of submarines. (Submarines are vulnerable vessels, actually, and can be put out of action easily if one only knows where they are. It is only the fact that they are hidden that makes them dangerous.)

Suppose a surface ship emitted a series of ultrasonic waves. There would be echoes from the sea bottom or even from schools of fish, but one might identify an echo sent back by a submarine and thus learn its direction and distance.

World War I ended before Langevin had perfected his device, but by World War II, such systems for echo location were working. The system was called *sound navigation and ranging*, where *ranging* is used in the nautical sense to mean "distance determination." Using the initial letters of the words, the system was called *sonar*.

Sonic Boom

An AIRPLANE traveling through air must push air molecules out of the way. This is possible because air molecules move so quickly that when they strike the surface of the plane they can almost always bounce away more rapidly than the plane is moving.

Sound travels through the air at a rate that depends on the natural motion of the molecules. This means that if an airplane travels at less than the speed of sound, the air molecules can bounce away. (The speed of sound is about 750 miles an hour. Smaller speeds are *subsonic*, from Latin words meaning "less than sound," and higher speeds are *supersonic*, or "more than sound.")

When the plane approaches the speed of sound, it is traveling as fast as the air molecules, which can't get away. More and more molecules are collected in front of the plane and a region of high pressure is set up. At first planes weren't properly designed to withstand the high pressure so that it seemed this might be an upper limit to their speed. With the proper design, however, the American pilot Charles E. Yeager managed to fly faster than sound on October 14, 1947.

If a plane *breaks the sound barrier* and flies at supersonic speed, it leaves the high-pressure region behind. If the plane slows up, the region races on ahead. In either case, the region manages to spread out and become air of ordinary pressures in time. If, however, the plane is close to the surface and its nose is aimed somewhat downward; the region of high pressure, like an enormous sound wave, will continue moving downward and will reach the surface before it is spread out. There will then be a *sonic boom*, which will sound like a loud bang. The air vibrations will rattle houses, break windows, and do considerable damage if strong enough. This was one of the objections to building the SST, or giant *supersonic transport* plane. Some people objected to the sonic booms it would be continually producing.

Space Station

The English scientist Sir Isaac Newton supplied the theory that explained the manner in which an object might be placed in orbit about Earth, but it was not until 1957 that sufficiently powerful rockets were built for the purpose. It was only then that adequately advanced computers for use in plotting the rocket orbits were developed, for that matter.

The initial satellites were relatively small, only large enough for some instruments. Eventually, rocket engines were made with adequate power to lift into orbit a capsule large enough to hold men. In 1969, a vehicle holding three men was fired to the moon. It retained sufficient fuel to land two of the men on the moon in a smaller vessel, and bring all three back to Earth.

Even so, these larger vessels could only serve as temporary homes for human beings. Both the United States and the Soviet Union have expressed interest in something more ambitious.

It would be possible to place a very large vehicle in orbit if, instead of attempting to launch it all at once, it were sent into space in parts. Men could be sent out also to put the individual parts together and, in the end, a large structure, capable of housing eighteen to thirty-six men, might be circling Earth at a height of two to three hundred miles above its surface.

Within the structure, men could carry out astronomical observations of Earth, the moon, the sun, and, indeed, of the universe, being unhampered by obscuring atmosphere. An observatory established on the moon might be larger and more comfortable, but it would also be more distant and harder to reach. The structure nearer to Earth is called a *Manned Orbital Laboratory*, abbreviated *MOL*, but the popular term for it is *space station*.

216

Stellarator

In the 1950s, physicists began to attempt to confine hydrogen gas at temperatures over a hundred million degrees, in order to initiate controlled nuclear fusion and liberate vast quantities of energy.

At first they tried to confine the gas in a straight cylindrical tube, holding it away from the walls (which would cool the hot gas and end the chance of fusion) by a magnetic field. This could be done, but only for a millionth of a second or so. Some arrangement had to be worked out which would produce a more stable confinement.

One possibility was to make the straight cylinder into a round hollow tube like a doughnut. This would eliminate the ends, which were the weak points of the earlier arrangement. Unfortunately, other problems showed up. The inner edge of the doughnut was shorter than the outer edge so that the magnetic field was more concentrated and therefore stronger along the inner edge. With an uneven magnetic field, the gas was forced to the outside of the hollow and made contact with the walls.

In 1951, the American astronomer Lyman Spitzer, Jr., suggested that the doughnut be twisted into a figure eight. In that way, the inner edge in one half of the figure becomes the outer edge in the other. As a result of the twist, one part of the tube crosses over another part so that the whole design carries the plasma through three dimensions.

Within such a tube, the hydrogen atoms could be accelerated to very high energies and, therefore, temperatures. The temperatures that were obtained were similar to those at the center of stars. The Latin words for "star" is *stella* and so the figure-eight tube was named a *stellarator*.

Although stellarators and other devices have not yet helped scientists achieve controlled fusion, the goal is coming nearer and nearer each year.

217

Strangeness Number

THERE ARE two force fields involved in nuclear reactions: the strong nuclear field and the weak nuclear field. Reactions that involve the former take place in a few trillionths of a trillionth of a second. Reactions that involve the latter take place much more slowly, taking as much as a billionth of a second or even longer.

In the 1950s, new particles were discovered, such as the kaons and the hyperons. These formed very rapidly, so that it was clear that they were affected by the strong nuclear field. It seemed that when they broke down, they should break down by way of the strong field as well, so that after being formed they would exist only for a few trillionths of a trillionth of a second, as is true of resonance particles.

The kaons and hyperons, however, broke down much more slowly through the weak nuclear field. Strange, thought the physicists, and so these particles came to be called *strange particles*.

In 1953, the American physicist Murray Gell-Mann suggested that in strong interactions, a certain quantity had to be conserved — that is, had to be kept unchanged. It was this quantity that involved the strange particles. In their breakdown, the quantity was *not* conserved, and hence they could not break down by way of strong interactions, only by way of the weak field, where the conservation was not necessary. This conserved quantity, in strong but not weak fields, was called the *strangeness number*.

The strangeness number for ordinary particles, such as protons, neutrons, and electrons, is zero, but for strange particles it is some positive or negative integer.

Later it turned out the strangeness number could be added to something called the *baryon number* to give a quantity that was easier to manipulate. This new quantity, since it applied chiefly to hyperons with their various electric charges, was called *hypercharge*.

218

Superconductivity

AN ELECTRIC CURRENT will flow through a metal wire; it will not flow through glass, sulfur, rubber, or many other substances. Metals, in general, conduct an electric current from one place to another. They are electrical *conductors* and have the property of *conductivity*.

The conductivity is never perfect. Every metal displays a certain *resistance* to the flow of the electric current. This resistance, acting like friction, turns some of the current to heat. If there weren't a battery or some other source of current continually working, the electric current would die out almost at once, with all its energy converted to heat. This would be true even in silver wires, though silver has the highest electrical conductivity known for any ordinary substance and the lowest resistance.

In general, vibrating atoms get in the way of the current. As temperature goes up, so that atoms vibrate more rapidly, resistance goes up. Conversely, as temperature goes down, resistance goes down.

Through the nineteenth century, scientists reached lower and lower temperatures. In 1908, the Dutch physicist Heike Kammerlingh Onnes reached a temperature of less than 4.2° above absolute zero (4.2 K.) and thus liquefied helium, the last substance to resist liquefaction.

Using liquid helium, he measured the electrical resistance of metals at very low temperatures. He expected this resistance to become low indeed but, in working with mercury, he found to his surprise that at 4.12° above zero (4.12° K.) the resistance dropped to zero. An electric current going through mercury below that temperature went on forever even with the source of electricity shut off. This was *superconductivity*. Other substances were found to be superconductive, too, and in 1968, an alloy of three metals was found to be superconductive at a record temperature high of 21° K.

219

Superfluidity

ONE OF THE MOST remarkable substances in the universe is helium. It is an absolutely inert gas and will not combine with anything. It is lighter than any gas but hydrogen, and it stays a gas at lower temperatures than any other known gas. It does not liquefy until a temperature of 4.2° above absolute zero (4.2° K.) is reached.

Liquid helium is even more remarkable. It does not freeze even at absolute zero, unless it is placed under pressure.

It is also remarkable for its ability to conduct heat. Heat is conducted with varying ease by different substances. Metals conduct heat more rapidly than nonmetals, and copper conducts it faster than any other metal. In 1935, though, the Dutch physicist William H. Keesom and his sister A. P. Keesom found that at temperatures below 2.2° K. liquid helium conducted heat with the speed of sound. Nothing else on Earth conducted heat so rapidly.

The Russian physicist Peter L. Kapitza found that the reason liquid helium conducted heat so well was that it flowed with unprecedented ease, carrying heat from one part of itself to another at least 200 times as rapidly as heat would travel through copper.

All gases and liquids have the ability to flow and they are therefore called *fluids*. The rate of flow is limited by the internal friction of molecules against molecules (viscosity) but in liquid helium there seems to be almost no viscosity at all. Helium not only has fluidity, it has *superfluidity*.

As a result of its superfluidity, liquid helium can leak through holes too small to allow gases to leak through. Something can be gas tight but not liquid-helium tight. The remarkable properties of helium below 2.2° K. are such that it is called *helium II*, to distinguish it from the more ordinary *helium I* above that temperature.

Synchrotron

In the 1930s, physicists developed methods for accelerating sub-atomic particles in order to give them high energies and send them smashing into atomic nuclei. The most successful of these was invented by the American physicist Ernest O. Lawrence in 1931. It whirled particles around and around, thanks to the driving force of a magnetic field, and it was therefore called a *cyclotron*.

By making larger and larger magnets, one could whirl the particles to greater and greater energies. The device only works well, however, if the mass of the particles doesn't change. As the particles go faster, their mass increases considerably (as Albert Einstein predicted they would in his special theory of relativity). This lowers the efficiency of the cyclotron and limits the energies it can produce.

In 1945, the Soviet physicist Vladimir I. Veksler and the American physicist Edwin M. McMillan, each independently, worked out a way to alter the strength of a magnetic field so as just to match the increase in mass. The two effects were *synchronized* (from Greek words meaning "same time") and the efficiency remained high. Such a modified cyclotron was called a *synchrocyclotron*.

In cyclotrons, the whirling particles spiral outward and eventually pass beyond the limits of the magnet. If the particles could be held in a tight circle, they could be whirled many more times before being released and still higher energies would be attained.

The English physicist Marcus L. E. Oliphant worked out a design for such a device in 1947, and in 1952, the first of the kind was built in Brookhaven National Laboratory on Long Island. It still made use of a synchronized increase in the strength of the field, but the spiraling of the particles, as in a cyclotron, was gone. The new device was therefore called simply a *synchrotron*.

Tachyon

THE SPECIAL THEORY of relativity, presented by Albert Einstein in 1905, says it is impossible for any particle with mass to reach or surpass the velocity of light in a vacuum (186,282 miles per second). On the other hand, particles of zero mass (like the photons that make up light) must travel through a vacuum at exactly 186,282 miles per second at all times.

According to relativity, any particle moving at faster-than-light velocities would have to have an imaginary mass. The mass would be some quantity multiplied by the square root of –1.

In 1962, a Russian-American physicist, Olexa-Myron Bilaniuk, and an Indian-American coworker, E. C. George Shidarshan, pointed out that particles with imaginary mass might exist without violating relativity. They would merely be required to move *always* at a speed faster than light. Such particles would slow down as they gained energy, but no matter how much energy they gained, they could never slow down to the velocity of light. That was the limit for these particles, as for ordinary particles, but from the other direction.

In 1967, the American physicist Gerald Feinberg popularized the notion and gave the faster-than-light particles the name *tachyons*, from a Greek word meaning "fast." Bilaniuk and Shidarshan then suggested *tardyon* (from a Latin word for "slow") as the name for all particles that travel more slowly than light. As for particles without mass that travel at exactly the speed of light (such as photons, neutrinos, and gravitons), Bilaniuk and Shidarshan suggested the name *luxons*, from the Latin word for "light."

Tachyons ought to leave a trail of light as they travel through a vacuum and therefore might be detected. However, they would move so fast that such flashes would last for unimaginably brief periods of time, and so far tachyons have not been detected.

222

Teflon

ATOMS OF CARBON have the unusual property of being able to join together in long chains, straight or branched, and in complicated systems of rings. Carbon atoms, so hooked up, can attach other atoms to themselves. In particular, they can be attached to hydrogen atoms, which are the smallest of all atoms and most easily fit into the angles where the carbon atoms form branches and rings. For this reason there are many hundreds of *hydrocarbons* that have been found or have been synthesized, and countless millions more can exist.

The only element with atoms small enough to substitute for hydrogen in this respect is fluorine. Fluorine is so hard to work with that *fluorocarbons* remained virtually unknown right down to modern times. The first example and the simplest, *carbon tetrafluoride* (CF_4) with one carbon atom and four fluorines, was prepared in 1926.

During World War II, *uranium hexafluoride* (UF_6) was used in connection with nuclear bomb research. Chemists began to study fluorocarbons in earnest. Fluorine atoms, it turned out, hold on to other atoms so strongly that they simply don't let go. Because they hang on as they do, they don't engage in chemical reactions, so that fluorocarbons are much more inert than hydrocarbons are. They don't even stick physically to other substances.

One of the fluorocarbons is *tetrafluoroethylene* ($CF_2 = CF_2$). Molecules of this compound can be made to hook together to form a long chain of carbon atoms, each with two fluorine atoms attached. The proper name for the chain would be *polytetrafluoroethylene*, the prefix coming from a Greek word for "many." The chemists who prepared the chain called it *Teflon* for short (a brand name), taking the letters from "tetrafluoro."

A thin coating of Teflon on frying pans will resist heat, and won't stick to food. For this reason, Teflon pans are easier to clean than ordinary ones and require little or no fat in frying.

Tektite

BACK IN THE eighteenth century, numbers of small, rounded pieces of greenish glass were discovered in what is now Czechoslovakia. In the last century, such pieces of glass have been found in other places: the Philippines, Australia, Texas, and the Ivory Coast. Some have been found on the sea bottom near those regions.

About 1900, an Austrian geologist, Eduard Suess, suggested they might be meteorites. Their composition was something like that of stony meteorites but they might have melted in their passage through the atmosphere and then cooled into a glassy substance. In fact, their shape was what would be expected of something moving rapidly through the air against atmospheric resistance.

This theory has come to be accepted. These objects are now thought to be meteorites that have partly melted in passage and so are called *tektites* (from a Greek word meaning "to melt"). The youngest ones, from the Far East, seem to have fallen only 700,000 years ago.

One suggestion is that the tektites originated on the moon. Large meteor strikes, like that which formed the crater Tycho, might have splashed lunar material upward fast enough to allow it to escape the moon's small gravitational field. Some of the material then moved far enough to be captured by Earth.

In 1969, scientists working on ways of keeping men alive under conditions in space decided it would be helpful to work on similar problems involving the undersea world. Four men spent two months fifty feet under the surface of the sea in small quarters designed to supply them with the necessities of life. This was called the *Tektite Project* because the work had been concerned with space to begin with and had ended under the sea, rather like the tektites themselves.

224

Terameter

THE METRIC SYSTEM, as first worked out by its French originators in 1795, had "kilo-" (from a Greek word meaning "thousand") as a prefix for its largest measurement. Thus, a *kilometer* is a thousand meters and a *kilogram* is a thousand grams. (The prefix is used so familiarly now even outside the metric system that a *kiloton* is a thousand tons and a *kilobuck* a thousand dollars.)

Scientists began to find it convenient to use prefixes for still larger measurements. They began to use the prefixes "myria-" and "mega-" for the purpose. These are from Greek words meaning "ten thousand" and "great," respectively. Thus, a *myriameter* is equal to ten thousand meters and a *megameter* to a million meters.

In 1958, there was international agreement to make the prefix "mega-" official and to add two more prefixes that would signify even larger measurements still. A thousand megameters would be a *gigameter*, from a Greek word meaning "giant," and a thousand gigameters would be a *terameter*, from a Greek word meaning "monster."

Such units could be convenient astronomically. For instance, the circumference of Earth is about 40 megameters and the moon is about 380 megameters from Earth.

Venus is about 42 gigameters from Earth at its closest, and the sun is 145 gigameters away. Jupiter, when at its farthest from Earth, is not quite a terameter distant. Pluto's orbit from end to end is not quite 12 terameters wide.

Even the largest prefix yet used, when applied to the meter, takes in only the solar system, which is a mere speck in the universe. In one year, light travels a little less than six trillion miles or nearly 10,000 terameters. This is one *light-year*, and even the nearest star is 4.3 light-years away.

Thermodynamics

THE STEAM ENGINE, developed and made practical in the eighteenth century, was quite inefficient. Very little of the fuel consumed in heating water and making steam was converted into useful work. Scientists grew interested in studying the way in which heat flowed from one point to another and the manner in which it was converted to work. They hoped to learn to make the steam engine more efficient and, perhaps, to understand the universe better. In this way a new science, *thermodynamics* (from Greek words meaning "heat movement") came into being.

Anything which could be converted into work was called *energy*. Heat, therefore, as well as certain other phenomena — such as light, sound, motion, and so on — came under that heading. By the 1840s, three different physicists, the Englishman James P. Joule and two Germans, Julius R. von Mayer and Hermann L. F. von Helmholtz, were convinced that if all the forms of energy were lumped together, the total remained constant in any closed system. Energy might be changed from one form to another but it could neither be created out of nothing, nor totally destroyed. This is called the *law of conservation of energy*.

It is also called the *first law of thermodynamics* because the new science could not be properly understood unless, to begin with, it was accepted that energy was conserved. Conservation of energy may be the most basic rule that scientists have yet discovered concerning the universe. No case has ever been found in which it doesn't hold.

On one occasion, there was serious doubt. In the 1890s, radioactivity was discovered and large quantities of energy seemed to be appearing out of nowhere within the atoms. But then, in 1905, the German-born physicist Albert Einstein showed that mass was a form of energy and that the huge energies produced by radioactivity were compensated for by the disappearance of tiny amounts of mass.

226

Thermosphere

It is clear that temperature drops steadily as one climbs mountains. Very high mountains have their peaks covered with snow even in the summer — even on the equator. In 1646, the French mathematician Blaise Pascal sent his brother-in-law up a mountainside with a barometer, and found that air pressure decreased with height. That meant that air grew less dense with height.

It is not surprising, then, that temperature drops with altitude, since the thinner air is, the less heat it can hold. In the 1890s, the French meteorologist Leon P. Teisserenc de Bort sent thermometers up in unmanned balloons and recorded a continuing drop of temperature. At a height of ten miles, it was down to –55° C.

The thinner the air, the less heat it can hold altogether. But as the air grows thinner, the atoms and molecules within it become fewer, and the heat striking it from the sun need be distributed among only those fewer atoms and molecules. Each atom or molecule has a larger share of heat, and the temperature (heat content per atom) tends to rise.

This latter effect becomes more marked the higher one goes. Up to a height of from five to ten miles, depending on latitude, the temperature does drop steadily. As one goes still higher, the second effect gains and for a while the temperature remains steady.

Eventually, at heights of beyond fifty miles, the temperature actually starts to go up, as revealed by rocket measurements in the 1950s. There continues to be less and less total heat, since there are so few atoms or molecules that even at high temperatures very little heat can be stored altogether. Nevertheless, by a height of 300 miles the temperature has reached 1000° C. This region of very thin air with high temperatures is sometimes called the *thermosphere* from Greek words meaning "heat sphere."

TIROS

THE FIRST ARTIFICIAL SATELLITES, launched in 1957, scarcely did more than show that the feat was possible. Then came satellites carrying instruments designed to study the upper atmosphere and the space just beyond. It was not long, however, before men began to use satellites in an attempt to study Earth itself.

We are right on the surface of Earth and can only see a small portion of it even when we rise as high as we can in an airplane or balloon. In order to understand the system of Earth's air circulation as a whole, it is necessary to take measurements in many places at many times and try to correlate the results. And yet it is only in certain parts of the world that such measurements are taken continually. Over vast stretches of sea and land, few or no measurements are taken. This adds to the difficulties of long-term weather forecasting.

But what if a satellite, circling Earth at a height of a couple of hundred miles from the surface, continually took pictures of the cloud cover? Information could be gathered in a few minutes that could not be gained in any other way. On April 1, 1960, a satellite was launched with two television-type cameras and with equipment for storing photographs and transmitting them to Earth on command. It was called *Television and Infrared Observation Satellite I* or, using the initials, *TIROS I.*

TIROS I was completely successful, as were other satellites of the same type. Thousand of pictures of Earth's cloud cover were taken and transmitted. The view of Earth with its spiraling clouds became familiar to all. Hurricanes were seen at their very start and were kept in view at all times. The overall changes of the cloud cover with the seasons were followed. It was clear that the satellite program was no mere stunt; it could be, and was, of important and practical use.

Tranquilizer

MEN HAVE ALWAYS looked for some way of inducing peace of mind and of making themselves feel better. They have used natural products for the purpose — smoked tobacco, chewed betel nuts and cocoa leaves, drunk fermented fruit juices and extracts of the poppy.

Modern chemistry invaded this area at the turn of the century. In 1882, a new organic chemical was discovered which was called *barbituric acid*. (The name was supposed to have arisen because the chemist who named it had a girl friend called Barbara.) In 1903, the German chemist Emil Fischer discovered that by attaching certain atom groupings to the barbituric acid molecule, he obtained a substance that had a soothing effect when swallowed. This and other similar compounds are grouped together as *barbiturates*.

The barbiturates, when taken in the proper dosages, do not relieve pain (they are not *analgesics*, from Greek words meaning "no pain") but they do seem to allay minor discomfort and anxiety and succeed in calming people. They are *sedatives* (from a Greek word meaning "to calm").

Barbiturates induce sleep, however. The larger the dose, the deeper the sleep, and the more difficult the arousal. *Sleeping pills*, if enough are taken, become a way of committing suicide.

In 1952, a drug was extracted from the dried roots of an Indian plant with the scientific name of *Rauwolfia serpentinum* (named for a sixteenth-century German botanist, Leonhard Rauwolf), which had been used as a calming influence by inhabitants of the regions in which it grew. The drug extracted was named *reserpine* (an abbreviation of the Latin name).

It was the first of the *tranquilizers*, drugs that reduced excitement and made a person tranquil without, however, inducing sleep or making him difficult to rouse.

229

Transduction

In the first third of the twentieth century, some biochemists were studying chromosomes in the cell nuclei, and others were studying viruses. Viruses are objects which are much smaller than cells — indeed, about the size of chromosomes. They don't seem to grow and multiply in the ordinary manner, but once they get inside a cell, they multiply very efficiently.

Both chromosomes and viruses were found to contain nucleic acid as a key component, and biochemists began to speculate that a virus might be a kind of independent chromosome.

A chromosome is made up of a chain of genes which are capable of directing the formation of certain protein molecules, making use of the complex chemical machinery of the cell to do so. A virus, when it gets into a cell, somehow takes over, making use of the cell machinery to form proteins necessary for its own purpose (rather than the cell's). It is a kind of intracellular parasitism.

Viruses, it would seem, can be so intimately involved with the cell machinery as to become a more or less permanent part of it. One or more genes of a virus may permanently join the genes in the cell's chromosomes, dividing when the rest of the chromosome divides and passing on into the daughter cells. This may explain the ability of some viruses to affect a host organism permanently after a single infection.

The phenomenon whereby a virus adds on to a chromosome and produces a permanent change in the characteristics of a cell — one that is carried on through generations — is *transduction*, from Latin words meaning "to lead across." Possibly, someday, biochemists will use viruses to add a desirable gene to a cell nucleus that lacks it. This is one of the techniques which may be used to change and modify the machinery of inheritance in a human being — the *genetic engineering* of the future.

Transfer RNA

THE STRUCTURE of the various molecules of deoxyribonucleic acid (DNA) in the chromosomes determines the structure of the enzymes produced in a particular cell. The DNA molecules, made up of chains of nucleotides, are in the nucleus, however, while the enzymes, made up of chains of amino acids, are in the cytoplasm.

Molecules of ribonucleic acid (RNA) pass from nucleus to cytoplasm. A particular RNA molecule is formed with a structure modeled on that of a particular DNA molecule. This *messenger RNA* travels into the cytoplasm, where it controls the formation of an enzyme molecule.

For this, every different combination of three nucleotides (a *codon*) along the chain of the RNA molecule must represent a particular amino acid — but how is that done?

In 1955, the American biochemist Mahlon B. Hoagland discovered small molecules of RNA (much smaller than messenger RNA) dissolved in the cell fluid of the cytoplasm. These came in a number of different varieties. Each variety had a particular combination of three nucleotides at one end, which would fit on one particular codon of the messenger RNA. At the other end of the small RNA molecule, an amino acid could be fitted, but only a particular kind.

The three particular nucleotides at one end of the small RNA always went along with one particular amino acid. This meant that when a whole series of these small molecules fitted itself onto the messenger RNA chain, the nature and order of those molecules were determined by the nature and order of the codons on the messenger RNA chain; the nature and order of the amino acids at the other end of the small molecules were also determined; and a particular enzyme molecule was formed.

Because the small molecules transferred the information from the messenger RNA to the enzyme, they were named *transfer RNA.*

231

Transistor

FOR ELECTRONIC DEVICES to work well, electric currents must be delicately controlled. They must be started, stopped, amplified, or altered, in a tiny fraction of a second. The first device that made this possible was an evacuated bulb called a *valve* in Great Britain and a *tube* in the United States. A heated filament within the tube released a flood of electrons that streamed across the vacuum within the tube. It was this flow of light electrons that could be easily and quickly controlled. Tubes, however, were bulky and fragile, and took time to heat.

The situation changed in the 1940s when an English-American physicist, William Shockley, and his coworkers studied semiconductors and their manner of working. The semiconductors existed in two types, some with a small surplus of electrons and some with a small deficit. In 1948, Shockley found that he could prepare combinations of these two types in such a way that the current passing through them could be controlled as conveniently and in the same manner as the electron flow through a vacuum.

Of course, the electron flow through a semiconductor did not progress with the nearly total absence of resistance that characterizes electron flow through a vacuum. In the semiconductors, the electrons are transferred across a resistance. John R. Pierce, who worked for the same company as Shockley's group, suggested that the new device for controlling electron flow be given a name taken from the phrase "*trans*-ferred across a re-*sist*-ance." It came to be called a *transistor*.

Where there had been tubes, there could now be tiny transistors. As a result, radios and other electronic equipment could be made much smaller than before, and more rugged. What's more, radios would start the moment they were turned on. Since there was no filament to heat, there was no warm-up period. Transistors also replaced the tubes in computers and in consequence, these were drastically reduced in size.

232

Transuranium Elements

In 1789, the German chemist Martin H. Klaproth discovered a new element which he named *uranium* after the planet Uranus, which had been discovered only eight years before.

In the first half of the nineteenth century, the relative masses (or *atomic weights*) of the atoms of the various elements were determined. It finally appeared that uranium had the most massive atoms of any known element. Although many more elements were discovered in following decades, uranium retained this pride of place. Its atomic weight (238) was higher than that of any other element even as late as 1940.

When the English physicist Henry G.-J. Moseley worked out a method for determining the electric charge on an atomic nucleus (the *atomic number*) it turned out that uranium had the highest of all — 92. By 1940, chemists were quite certain that uranium atoms were the most complex that occurred naturally on Earth. If more complex atoms ever existed, they were probably so radioactive that they broke down and were gone eons before man appeared on Earth. (Uranium itself is radioactive also, but breaks down very slowly.)

Might not more complex atoms be made in the laboratory, though? In 1934, the Italian-born physicist Enrico Fermi bombarded uranium atoms with neutrons, hoping to form element 93. For a while he thought he had succeeded and called the new element *Uranium X*. He was, however, causing the uranium atom to split (uranium fission) and this confused matters.

With Fermi switching to the study of uranium fission, it was left to two American-born physicists, Edwin M. McMillan and Philip H. Abelson, to isolate element 93 in 1940. They named it *neptunium* after Neptune, the planet beyond Uranus. Neptunium was the first of the *transuranium elements* (beyond uranium). Since 1940, no fewer than thirteen more, up to element 105, have been prepared in the laboratory.

Tsunami

WHEN AN EARTHQUAKE takes place on one of the continents, it can do a great deal of damage. It might seem that a quake far out at sea, however, could be ignored. The water would shake a bit, but surely no one would be hurt. And yet an earthquake at sea can be more dreadful than one on land.

The seaquake will set up a wave that is not very high in mid ocean but stretches across the surface for an enormous distance and therefore involves a large volume of water. Such a wave spreads outward in all directions from the point at which the quake took place. As it approaches land and as the ocean gets shallower, the stretch of water in the wave is compressed front and rear and piles up higher, then much higher. If the wave moves into a narrowing harbor, its volume is forced still higher, sometimes fifty to one hundred feet high.

That tower of water, coming suddenly and without warning, can break over a city, drowning thousands. Before the wave comes in, the preceding trough arrives. The water sucks far out, like an enormous low tide, and then the wave comes in like a colossal high tide. Because of this out-and-in effect, the huge wave has been called a *tidal wave*. This is a poor name, though, for it has nothing to do with the tides.

In recent decades, the name *tsunami* has been used more and more frequently. This is a Japanese word meaning "harbor wave," which is an accurate description. The Japanese, living on an island near the edge of our largest ocean, have suffered a great deal from tsunamis.

The largest in recent years was in 1883, when the volcanic island Krakatoa in the East Indies exploded and sent hundred-foot tsunamis crashing into nearby shores. About B.C. 1400, an Aegean island exploded and a tsunami destroyed the civilization on the nearby island of Crete. Still a third famous tsunami destroyed the city of Lisbon in 1755.

UFO

On June 24, 1947, a Seattle businessman, Kenneth Arnold, saw from a plane, in the vicinity of Mount Rainier, a series of shining disklike objects moving through the air in a skipping fashion. He reported this, and from his description, men began to speak of *flying saucers*.

Since Arnold's first sighting, many thousands of reports of such things have appeared. Some were easily explained, proving to be anything from mirages to the sightings of bright planets. Others have turned out to be hoaxes. And some, generally because of insufficient information, remain unexplained.

Nevertheless, a veritable hysteria seized some people who were sure that the objects were enemy planes or even extraterrestrial spaceships manned by otherworld life forms. The American government and sober scientists were accused of hiding the facts, and people who tended to see conspiracies everywhere were never-ending in their ridiculous theories and accusations.

Hounded onward by people ranging from well-meaning unsophisticates to fiery-eyed crackpots, the government instituted investigation after investigation, finding nothing and offering only fresh fuel for cries of conspiracy. Since the term *flying saucer*, though still popular, was clearly inadequate, investigators called the things seen in the air *unidentified flying objects*, as objective a term as possible. This was quickly abbreviated to *UFO*, sometimes pronounced as initials, sometimes as a word, and flying saucer enthusiasts or investigators are jocularly called *ufologists*.

Undoubtedly, there remain phenomena in the atmosphere that are as yet poorly understood by scientists, but the ufologist's "understanding" that the mysteries are to be explained by flights from other worlds is unlikely in the extreme.

Ultracentrifuge

Most protein molecules have weights from thousands to millions of times as great as that of the hydrogen atom. The ordinary methods for determining molecular weights fail in the face of such size.

Protein molecules are so large, in fact, that the endless motions of the water molecules surrounding them when they are in solution barely suffice to keep them distributed evenly. If they were somewhat larger, they might actually settle out of solution under the pull of gravity. Naturally, the larger they are, the more rapidly they settle out and in that way one can determine their molecular weight — by the rapidity with which the settling takes place.

When anything settles out of solution, it forms a *sediment* (from a Latin word meaning "settling"). The rate at which proteins settle out is therefore the *sedimentation rate*.

In order for actual proteins to have a measurable sedimentation rate, the force of gravity would have to be greater than it is. It is impossible to arrange this condition, but there is something which can be used to imitate gravity. If a container is whirled rapidly, its contents are forced away from the center of rotation. This is the *centrifugal effect* (from Latin words meaning "to flee from the center"), and the instrument is a *centrifuge*.

In 1923, the Swedish chemist Theodor Svedberg developed an *ultracentrifuge* (beyond the centrifuge) that whirled so rapidly it produced an effect large enough to force protein molecules to move outward through the water solution. A sedimentation rate was obtained which could be used to determine the molecular weight of proteins.

Nowadays, ultracentrifuges are used which whirl at 75,000 times each minute and produce centrifugal effects that are 400,000 times as powerful as gravity.

Uncertainty Principle

In 1900, the German physicist Max K. E. L. Planck demonstrated that energy existed in tiny separate bits. Each bit was called a *quantum* (a Latin word meaning "how much?"), a term first introduced by Albert Einstein in 1905. The amount of energy in one quantum depended on the wavelength of the radiation associated with the energy. To calculate the energy per quantum from the wavelength of the radiation, one had to use a very small number called *Planck's constant*, and it turned out that this number set important limits to certain phenomena.

For instance, it had always been assumed that scientists could make measurements that were as accurate as they wished. If they constructed tools that were perfect, then the measurements would be perfect, too, and would have no uncertainty about them.

In 1927, however, a German physicist, Werner Heisenberg, challenged this assumption. Measurements always involved the use of energy in one fashion or another (even if only light to see by), and the energy had to come in certain sizes dictated by Planck's constant. If the measurement one tried to make was so small that the particle of energy used was large in comparison, then the measurement would be uncertain no matter how carefully it was made or how perfect the measuring instrument. (It would be like playing the piano with boxing gloves on. Even the most perfect pianist couldn't do a good job.)

Heisenberg calculated in particular that if one tried to measure the position of an object with greater and greater accuracy, then the simultaneous measure of its momentum (its mass times its velocity) would be determined with less and less accuracy, and vice versa. The uncertainty of one measurement multiplied by the uncertainty of the other had to be no smaller than a certain fraction of Planck's constant. This was the *uncertainty principle* and it is a limit set by the structure of the universe.

Van Allen Belts

On October 4, 1957, the age of space opened when the Soviet Union sent *Sputnik I,* the first of the artificial satellites, into orbit. The United States was not far behind. Its first artificial satellite, *Explorer I,* was sent into orbit on January 31, 1958.

Of course, satellites were not merely intended to go into orbit to prove the strength of their rocket engines. They carried instruments designed to make measurements of various kinds. One obvious type of measurement was that of counting the number of cosmic ray particles and other energetic particles in the regions through which the satellite passed. *Explorer I* was designed to do that.

As *Explorer I* rose higher, it recorded more and more energetic particles, then went dead and reported nothing. *Explorer III* was launched in March 1958 with a more rugged counter. That, too, went dead.

It occurred to the American physicist James A. Van Allen, who was in charge of this part of the work, that the trouble might be not that there were no particles above a certain level, but that there were too many. Perhaps the counters were flooded and couldn't work.

When *Explorer IV* was launched on July 26, 1958, it contained a counter with a lead shield that would let through only a small portion of the particles (like a man wearing dark glasses to protect his eyes from glare). This time the results were conclusive. The number of energetic particles was indeed high, far higher than anyone had imagined.

Apparently, energetic particles, originating for the most part from the sun, were trapped in Earth's magnetic field and existed in a thick belt outside the atmosphere all around Earth. On closer study, there seemed two or even three belts surrounding Earth, and these were called the *Van Allen belts* at once. In later years, to adjust the name to those of other portions of the atmosphere, the region came to be called the *magnetosphere.*

Vasopressin

IN THE EARLY 1940s, it was found that the pituitary gland produced a substance that controlled reabsorption of water passing through the kidneys. By such control, the body was kept from losing too much water through urination.

An extract from the pituitary, called *pituitrin*, was found to contain the hormone. Any factor which increases urine volume is said to be *diuretic* from a Greek word meaning "to urinate." Pituitrin, by promoting reabsorption of water, lessened the need to urinate and was therefore the opposite of diuretic. It was said to contain an *antidiuretic hormone*.

Pituitrin possessed two more important abilities. It increased blood pressure by bringing about the contraction of blood vessels. This was called *vasopressor* activity from Latin words meaning "vessel compressing." It also contracted the muscles of the pregnant uterus in preparation for birth. This was called the *oxytocic effect*, from Greek words meaning "quick birth."

The American biochemist Vincent du Vigneaud obtained two pure substances from pituitrin in the early 1950s. Since one showed the vasopressor effect and the other the oxytocic effect, he called them *vasopressin* and *oxytocin*, respectively. Vasopressin also produced the antidiuretic effect, so it was also the antidiuretic hormone.

Du Vigneaud analyzed the two hormones and showed that each was made up of eight amino acids, which made them very simple proteins indeed. He identified the eight amino acids and worked out the order in which they appeared in the molecule. Then, in 1955, he synthesized the hormones from the individual amino acids, which he put together in just the right order. He showed that the synthetic substances had all the properties of the natural hormones. It was the first time any natural protein (albeit a simple one) had ever been synthesized.

Venera

THE NEAREST SIZABLE heavenly body to us is the moon, which is a mere quarter of a million miles away. Therefore, man's ventures into space have naturally concentrated on the moon, more than on any other extraterrestrial body. In July 1969, men actually landed on the moon, and there have been additional trips there since then.

The next nearest bodies are the planets Venus and Mars. Venus can be as close to us as twenty-five million miles and Mars thirty-five million. Of the two, Mars is the less inhospitable, so that it is the next target for a manned vessel. Both, however, have been the target for unmanned probes.

The Soviet Union was not satisfied to have these probes merely fly near Venus. Soviet scientists have labored to design rockets that would make a soft landing on the planet, and then radio back information from within the atmosphere. Those rockets intended for this purpose they called *Venera*, after the name of the planet.

The first attempts, *Venera IV, V,* and *VI,* were sufficiently accurately aimed to strike Venus' atmosphere. The instrument package began descending through it by parachute. In each case, however, the atmospheric conditions were so extreme as to cause the transmitting devices to break down before the package had descended to within a dozen miles of the surface.

Finally, on December 15, 1970, *Venera VII* succeeded. It made a soft landing on Venus, the first time such a feat had been accomplished on any world in space other than the moon. The data sent back by *Venera VII* showed the temperature on Venus' surface (at its point of landing) to be 474° C., over a hundred degrees higher than is required to melt lead. The atmospheric pressure is about 1300 pounds per square inch, or ninety times that of Earth's atmosphere. Considering that Venus' atmosphere is almost entirely carbon dioxide, the planet seems inhospitable indeed.

W-Boson

MOST SUBATOMIC PARTICLES act as though they are spinning. The spin gives the particle a property called *angular momentum* and physicists can measure this. They use a system of measurement whereby some particles end with a spin expressed as a whole number. A photon, for instance, has a spin of 1. Others have spins expressed as a *half-number*. Electrons and protons have spins of $\frac{1}{2}$.

Particles are distributed among different energy levels, and this distribution can be worked out by either of two kinds of statistical methods. One was worked out by the Italian-born physicist Enrico Fermi and the English physicist Paul A. M. Dirac. Such *Fermi-Dirac statistics* hold for particles with spins of half-values. Particles like the electron and proton are therefore called *fermions*.

The other statistical method was worked out by the Indian physicist Satyenda N. Bose and by the German-born physicist Albert Einstein. The *Bose-Einstein statistics* hold for particles with whole-number spins and these are called *bosons*. The photon is a boson.

The photon is involved in electromagnetic interactions. Other kinds of interactions are also associated with special particles. The strong interaction associated with atomic nuclei involves the pion, which has a spin of 0 and is also a boson. Gravitational forces involve gravitons, which have spins of 0 and are bosons.

The only remaining interaction is the weak interaction associated with events such as radioactivity. Physicists feel this too must be associated with a particle, one that may be particularly massive, even more massive than a proton. Theory predicts it would have a spin of 1, so it would be a boson, too. It is distinguished from other bosons by being called a *W-boson*, where the *W* stands for *weak interaction*. The W-boson has not yet been detected.

241

Weak Interaction

In 1935, the Japanese physicist Hideki Yukawa worked out the manner in which the atomic nucleus, made up of protons and neutrons, held together. There was a *nuclear interaction*, setting up an attraction among the particles, and that proved to be the strongest known force, over a hundred times as strong as the electromagnetic interaction and trillions upon trillions of times as strong as the gravitational interaction.

The nuclear interaction is so strong that once two particles are close enough to allow the interaction, the result must follow in an unimaginably short period of time. Thus, once a pion is in the neighborhood of an atomic nucleus, a reaction follows in the space of ten trillionths of a trillionth of a second.

It would seem that any subatomic particle capable of being involved in the nuclear interaction would break up in that period of time. There are strong theoretical reasons for supposing so, and yet this does not happen. The pion, for instance, breaks down after a few billionths of a second. This seems short enough, but the pion is more than a hundred trillion times as long-lived as it is "supposed" to be. A neutron can exist for ten minutes and more without breaking down.

Apparently, there must be another kind of interaction involving subatomic particles, much weaker than the one Yukawa had discovered. As early as 1934, the Italian-born physicist Enrico Fermi had worked out the theory behind such a *weak interaction* and Yukawa's came to be called the *strong interaction* in contrast. The weak interaction is much weaker even than the electromagnetic interaction, but is still trillions of times stronger than the gravitational interaction.

There are some subatomic particles, like the electron and the neutrino, that are involved only in weak interactions and that is why most ordinary radioactive breakdowns take place as slowly as they do.

Whiskers, Crystal

IN CRYSTALS, the component parts — atoms, ions, or molecules — make up an orderly array. The orderliness makes possible a strong cohesiveness that would not be possible if the various atoms, ions, or molecules were arranged in a random heap. Still, when chemists calculated how strong a crystal ought to be on the basis of this orderly arrangement, it always turned out that the crystals in actual fact were far less strong than was to be expected.

When chemists learned to study the structure of crystals in detail, by x-ray diffraction, it turned out that the orderliness was never perfect. There were invariably small regions of disorder where there was a missing atom or a superfluous one, where lines of atoms didn't meet exactly or did so at a slight angle. When stress was placed on a crystal, it gave at one of these points of weakness, long before the orderly portions were affected. Once a tiny crack appeared at these critical points, it spread rapidly.

If a crystal could be formed free of all such defects, it would be much stronger than ordinary crystals of that type. It had long been known that when crystals were slowly formed from solution, tiny hairlike projections appeared on the crystals. These are called *whiskers* because that's what they looked like. In the 1950s, these whiskers were studied and found to be perfect crystals and therefore much stronger than ordinary crystals of the same size. Carbon whiskers have been found to have a great resistance to being pulled apart (*tensile strength*, from a Latin word meaning "to be stretched"). The tensile strength of carbon whiskers is from fifteen to seventy times that of ordinary steel.

In 1968, Soviet scientists produced an ordinary, but very small, crystal of tungsten without imperfections. It could bear a load eight times as large a piece of steel the same size could carry.

Wolf Trap

By the mid twentieth century, there seemed little hope, if any, for advanced life forms anywhere in the solar system, except on Earth itself.

But what about very simple forms of life, which are much hardier than complex animal forms? There are certain seasonal changes on the surface of Mars that might possibly be due to plant growth. Biologists have set up chambers containing the equivalent of a possible Martian atmosphere, maintained at a Martian temperature, and simple plants have managed to live and grow in it. It even seemed possible that bacteria might manage to grow in secluded places on the moon where bits of air and water might have persisted under the immediate surface.

In recent years, hopes have weakened. Men have actually landed on the moon and brought back samples of rocks — but these have shown no signs of organic matter. (To be sure, very little of the moon has yet been sampled and no matter from beneath the surface has been brought back.)

The Martian atmosphere has been found to be even thinner and dryer than had been thought. Still, hope has not vanished entirely with respect to Mars. Men won't set foot on Mars for years yet, but possibly some unmanned device could send back information. One possible gadget was devised by an American biologist, Wolf Vishniac, in 1960. An instrument would be soft-landed on Mars. It would suck a sample of neighboring soil or dust into a clear solution containing chemicals on which life could live. If life forms were included in the soil, they might grow and increase in numbers, making the solution cloudy or increasing the acidity. In either case, information to that effect would be relayed back to Earth. Since the device is designed to trap any primitive life present, it seemed appropriate to make use of the first name of the designer and to call it a *Wolf trap*.

Xerography

EVER SINCE WRITING was invented five thousand years ago, people have been interested in copying what was written. For almost all of this time, copying has been possible only painstakingly, by hand, and each copy was as hard to do as the original had been.

Once printing was invented, any number of copies could be prepared, but to do that required a printing press, a lot of type, and considerable skill in setting up type. A mimeograph machine is much more of a small-scale operation, but this must use liquid ink and can be messy. Carbon paper is dry and prepares copies as one types, but only a few at a time.

What if one uses carbon powder instead of ink and lets electrostatic attraction do the work? Suppose a sheet of white paper is electrically charged. The charge would attract any fine particles of carbon that might be present and the entire sheet would be covered with a thin layer of carbon. Light, impinging on the paper, however, would cause it to lose the charge.

But suppose light shines through a paper with print on it and strikes the charged paper. Everywhere, except where the print casts a shadow, the charge is lost. Carbon powder clings only to the shadow. The paper is heated so that the powder (which contains a resin to make it stick) clings permanently. The second paper is then a copy of the first, and many copies can be made rapidly.

The process is called *xerography*, from Greek words meaning "dry writing," since nothing wet is used.

By 1960, the American inventor Chester Carlson had made the method practical enough for eventual use in almost every office, and this has revolutionized office procedure. The system most familiar to the general public has the trade name *Xerox*, from "xerography."

245

X-Ray Stars

UNTIL THE 1950s, the only radiation we could detect from the sky was that of visible light, cosmic rays, and certain radio waves. Other kinds of radiation were mostly absorbed by the atmosphere.

In the 1950s, however, rockets carried instruments beyond the atmosphere and other radiations were detected. For instance, it was found that the sun radiated x rays, which was very surprising since it was not thought hot enough to do so. It was discovered, though, that the sun's corona was at a temperature of one or two million degrees and this was hot enough for x-ray emission.

The Italian-American physicist Bruno Rossi wondered if the solar x rays might hit the moon and be reflected. In 1962, rockets were sent up with instruments designed to detect any x rays coming from the night sky. They detected such radiation, but not from the moon. It was coming from the general direction of the center of the Galaxy. Objects far beyond the solar system seemed to be radiating x rays so strongly as to produce measurable quantities even across light-years of space.

Beginning in 1963, the American astronomer Herbert Friedman began a program of rocket detection designed to cover the sky in order to locate regions particularly rich in x-ray emission. A number of such regions were located and these were called *x-ray stars*.

The nature of the x-ray stars proved puzzling, but one object which definitely radiates x rays is the Crab Nebula. This is the remains of a gigantic stellar explosion, the effects of which reached Earth a little over 900 years ago. Even after 900 years, the energy of the explosion has not faded out and the nebula is a rich source of radio waves as well as x rays. It even emits cosmic rays, the most energetic radiation of all. In addition, the nebula contains a pulsar, and this may be responsible for at least part of the radiation.

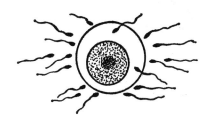

Y-Chromosome

EVERY SPECIES of animal has a characteristic number of chromosome pairs in its cells. In the case of the human, that number is 23.

In females, each pair consists of two chromosomes that appear exactly alike. Not so in the male. After most of the chromosomes have been paired off, two chromosomes are left that are distinctly different. One is rather longer than average, the other is a mere stub. The long chromosome is called the *X-chromosome* and the stub is the *Y-chromosome*. In the female, there are two X-chromosomes. The X- and Y-chromosomes, which differ in arrangement in the two sexes, are called the *sex chromosomes*.

Egg cells and sperm cells, when formed, contain only twenty-three chromosomes, one of each pair. Since the female has two X-chromosomes (XX), all the egg cells have one X. Since the male has an X-chromosome and a Y-chromosome (XY), half the sperm cells end up with an X and half with a Y. It is just random chance that determines whether an X-sperm or a Y-sperm fertilizes a particular egg cell. If the former, the result is an XX fertilized egg, which develops into a female; if the latter, an XY fertilized egg, which develops into a male.

Chromosome assignment is not always perfect in the formation of egg cells and sperm cells (which together may be called *sex cells*). Sex cells with abnormal numbers of chromosomes usually don't form a fertilized egg that can develop fully. In the late 1960s, however, it was discovered that sex cells with abnormalities in their sex chromosomes might form embryos that would develop and give rise to individuals with XXX in their cells, XXY, XYY, and so on. XYY individuals have been found to be males, generally tall and strong, and often given to irrational fits of violence. Society, in its treatment of offenders, must now take into account their chromosomal make-up in determining the extent to which they can be held responsible for their actions.

Zinjanthropus

AFTER THE ENGLISH NATURALIST Charles Darwin had published his theory of evolution in 1859, there was naturally a search on for fossils of the apelike precursors of mankind. In 1891, a Dutch paleontologist, Eugene Dubois, discovered fossil remnants in Java of a manlike being with a skull capacity distinctly less than that of any normal modern man. Others were eventually found in China and Africa and the history of man and his immediate ancestors was moved back a million years.

Yet it seemed that the true origin of man must be searched even farther back in time. Two paleontologists, the Kenya-born Englishman Louis S. B. Leakey and his wife, Mary, were particularly interested in the Olduvai Gorge in Tanzania. This was a deep cut in the earth that exposed old layers which showed signs of being rich in fossils. Perhaps some might be prehuman. They searched meticulously, even using tweezers and small brushes.

Finally, on July 17, 1959, Mary Leakey crowned a more-than-a-quarter-century search by discovering fragments of a skull which, when pieced together, proved to encase the smallest brain of any manlike creature yet discovered. East Africa had been dominated by Arab traders before the coming of Europeans and the Arabic word for "East Africa" was *Zinj*. The Leakeys therefore called the new fossil find *Zinjanthropus* ("East African man," the last part of the word being Greek).

Zinjanthropus was advanced enough to form and use primitive tools, though the rocks in which the fossil was found were nearly two million years old. Zinjanthropus does not seem to be a direct ancestor of modern man, but in 1961, Louis Leakey discovered another fossil, slightly older than Zinjanthropus, which may be in our direct line of ancestry. This new fossil he called *Homo habilis* (Latin words meaning "skillful man"), for it already seemed to have hands as nimble and skillful as those of modern man.

Zone Refining

WITH ADVANCES in modern technology, it has become more and more important to deal with extremely pure materials. In preparing transistors, for instance, even a very few atoms of impurities are liable to give off or absorb electrons in such a way as to disrupt the delicate electron movements on which transistor properties depend. Impurities must often be kept down to only a couple of atoms per billion.

The traditional way of freeing a material from impurities is to dissolve it and then subject it to some chemical treatment that will cause it or the impurities, but not both, to crystallize out of solution. This is done over and over, and at each step more impurity is removed. This *fractional crystallization* is, however, a tedious process.

A much simpler process was worked out when it was realized that an impurity might be more soluble in the liquid form of a material than in the solid form, or vice versa. Imagine a metal rod which can be heated at some particular section or zone. That zone is brought just to the melting point. The heating process is moved along the ingot so that the melted portion travels. In no zone does the metal stay molten long enough to drip.

As the melted zone travels along the ingot, it collects impurities that dissolve more easily in the liquid than in the solid. By the time it has traveled from one end to the other, the impurities are concentrated in that end. A second zone of melting follows, and another, each one flushing out more of the impurities and forcing them to the far end. If the impurities are more soluble in the solid, the melted zone forces the impurities backward so that they collect in the near end. In either case, the impure end is cut off and what is left of the ingot is extremely pure.

This process of *zone refining* adds no chemicals to the ingot and so does not introduce new impurities.

ZPG

ONE MEASURE of man's success in the world is his number. There are now an estimated 3,600,000,000 people in the world, four times as many as there were less than two centuries ago.

Mankind's population has been steadily increasing since prehistoric times, as his restless brain and nimble hands steadily increased his control over the environment. The taming of fire, the herding of animals, the development of agriculture, metallurgy, writing — each contributed to the ability of mankind to multiply.

The steady population increase was accelerated in the nineteenth century because new lands were brought under large-scale agricultural production, because industrialization developed and spread, and because medical advances drastically reduced the death rate.

As a result, the time it takes for Earth's human population to double has steadily decreased. Now, with life expectancy at the seventy-year mark even in many regions which, while not industrialized, have taken advantage of modern notions of hygiene and medicine, man's population is expected to double in thirty-five years.

Increasing population and developing technology, however, have put an ever greater strain on Earth's resources and on its capacity to absorb industrial wastes. We seem to be corroding the quality of the environment faster than we can repair the damage.

Many feel that we must reorder our technology to make its first priority preservation of the environment and that we must call a halt to the continuing wild increase in human population. It is thought that there must be *zero population growth* (usually abbreviated to ZPG) if we are to survive. If this is so, then scientists face a vital problem. How are the physical, psychological, and sociological sciences to be used in such a way as to preserve the environment and halt man's population increase in an efficient and humane way?

Index

C

Calcitonin, 39
Calcium tungstate, 175
Californium, 128
Calvin, Melvin, 18, 176
Cambrian era, 67
Cancer, 153
Cantor, Georg, 7
Čapek, Karel, 201
Carbon, 4, 34, 41, 195, 211, 223
  meteorites and, 40
Carbonaceous chondrites, 40
Carbon dioxide, 18
  greenhouse effect and, 105
Carbon paper, 245
Carbon powder, 245
Carbon tetrafluoride, 223
Carbon whiskers, 243
Carborane, 41
Carcinogen, 153
Carnot, Nicolas L. S., 83
Carrington, Richard C., 213
Cash register, 73
Castor, 95
Cathode rays, 36, 129
Caton, Richard, 80
Cells, 49, 147, 149, 165
  chromosomes in, 124, 247
  retinal, 199
  structure of, 200
Centrifugal effect, 236
Centrifuge, 236
Cent XR–2, 206
Cepheids, 42
Cepheus, 42
Ceramics, 44
Čerenkov, Pavel A., 43
Čerenkov counters, 43
Čerenkov radiation, 43
Cermets, 44
Chadwick, Sir James, 23, 157, 161
Challenger, 24, 107
Chaos, 61
Charge, electric, 150
Charge conjugation, 63
Chemistry, 97

Chitin, 122
Chiu, Hong-Yee, 193
Chlorine, 155
Chlorophyll, 45, 176
Chloroplasts, 45
Choline, 2
Cholinesterase, 2, 155
Chondrites, 40
Chondrules, 40
Christofilos, Nicholas, 136
Christofilos effect, 136
Chromosomes, 47, 96, 230
  human, 124
  pairs of, 247
Circadian rhythm, 46
  pineal gland and, 180
Circle, 177
Cis trans configuration, 47
Cistron, 47
Citric acid, 126
Citric acid cycle, 126
Clairvoyance, 170
Clarke, Arthur C., 102
Claude, Georges, 175
Clausius, Rudolf J. E., 83
Clay, 44
Clock paradox, 48
Clone, 49
Clostridium botulinum, 35
Cloud chamber, 38, 50, 152
Clouds, 51, 212
Cloud seeding, 51
Clusters, galactic, 133
Coacervate, 52
Coal, 34
Cobalt, 92
Codon, 53, 231
Coelacanth, 54
Coenzyme, 55
Cohen, Paul J., 7
Coherent light, 127
Colchicine, 153
Cold light, 81
Coleridge, Samuel T., 138
Combustion, 212
Communication, 174
Communications satellites, 102

253

Proteins (*continued*)

laboratory synthesis of, 190, 208
molecular weight of, 236

Proterozoic era, 67

Protons, 11, 12, 23, 87, 108, 142,
156, 242

mass of, 113
spin of, 241

Protoplasm, 147

Pulsar, 32, 158, 191, 246

Pulse, 10

Q

Quantum, 237
Quantum theory, 164
Quark, 192
Quartz, 178
Quasar, 83, 103, 193, 196
nature of, 209
Quasi-mammals, 194
Quasi-stellar objects, 193

R

Radar, 145
Radiation, electromagnetic, 150
Radioactivity, 109, 198, 226
Radio astronomy, 118, 145, 193
Radiocarbon dating, 195
Radio telescopes, 118, 191
Radio waves, 27, 145
Radium, 109
Radon, 160
Ramsay, Sir William, 160
Ranger VII, 137
Rare earths, 3
Rauwolf, Leonhard, 229
Rauwolfia serpentinum, 229
Rayleigh, Lord, 160
Real number, 114
Red blood corpuscles, 210
Red dwarfs, 115
Red giants, 115
Red shift, 193, 196
Refraction, 184
Refrigerators, 64, 65

Reines, Frederick, 156
Relativity, theory of, 48, 61, 103,
131, 151, 167, 221, 222
REM sleep, 197
Reserpine, 229
Resistance, electrical, 219
Resonance, 198
Resonance particles, 198, 218
Retina, 30, 199
Retinene, 199
Reverberation, 9
Rhine, Joseph B., 170
Rhodopsin, 199
Ribonucleic acid (RNA), 143, 200,
231
Ribose, 69, 143
Ribosome, 165, 200
Rice, 100
Rime of the Ancient Mariner, 138
RNA, 143, 200, 231
Robot, 30, 201
Rochester, George D., 113
Rocket, 16, 117, 188
Rocket fuel, 33, 188
Rods, 199
Roentgen, Wilhelm C., 36
Rossi, Bruno B., 246
Rubber, 79
Ruben, S., 195
Rutherford, Ernest, 23, 128, 157,
161, 164, 204
Rutherfordium, 128

S

Satellites: artificial, 29, 62, 102, 137,
138, 139, 141, 159, 187, 216,
228, 238
natural, 119
Saturn, 119
Scandium, 3
Schaefer, Vincent J., 51
Schiaparelli, Giovanni V., 88
Schizophrenia, 202
Schmidt, Bernard, 203
Schmidt, Maarten, 193
Schmidt camera, 203

Ultrasonic sound, 176
Uncertainty principle, 237
Unconformity, 67
Ungar, Georges, 205
Unidentified flying object (UFO),
    235
Universe, 78, 88
    origin of, 27, 58, 60, 61
Upatnieks, Juris, 111
Uranium, 21, 109, 178
    breakdown of, 195
    discovery of, 233
Uranium hexafluoride, 223
Uranium X, 233
Uranus, 233

V

Vagusstoff, 2
Van Allen, James A., 136, 238
Van Allen belts, 136, 238
Van de Kamp, Peter, 22
Vasopressin, 239
Veksler, Vladimir I., 221
Venera, 240
Venus, 105, 240
Venus probe, 105, 138
Verne, Jules, 62
Vesicant, 155
Violet shift, 196
Virus, 116, 230
Viscosity, 220
Vishniac, Wolf, 244
Vision, 199
Visual purple, 199
Vitamin A, 199
Vitamins, 20, 55
Von Euler, Ulf S., 189
Von Helmholtz, Hermann L. F.,
    226
Von Koenigswald, Gustav H. R.,
    101
Von Leibnitz, Gottfried, 73
Von Mayer, Julius R., 226
Vonnegut, Bernard, 51

Von Neumann, John, 94
Von Sachs, Julius, 45
Von Weizsäcker, Carl, 78
V-particle, 123

W

Wald, George, 199
Walsh, Don, 25
Wastes, 185
Water, 186
Wave mechanics, 164
W-boson, 87, 241
Weak nuclear interaction, 104, 108,
    171, 218, 241, 242
Weather forecasting, 121
Weather satellites, 159, 228
Weber, Joseph, 104
Wegener, Alfred L., 57
Weiss, Pierre E., 75
Went, Frits W., 19
Wheat, 106
Whiskers, crystal, 243
White dwarf, 158
Whittle, Frank, 120
Wiener, Norbert, 68
Wigglesworth, Vincent B., 122
Williams, Carroll, 122
Wilson, Charles T. R., 38, 50
Wilson, Robert W., 27
Wind, 121
    solar, 213
Windsor, Duke of, 10
Wolf trap, 244
Writing, 245

X

X-chromosome, 247
Xenobiology, 88
Xenon, 160
Xerography, 245

Xerox, 245
Xi particle, 113, 163
X rays, 36, 246
   mutation and, 153
X-ray stars, 206, 246

Y

Yang, Chen Ning, 63, 171
Y-chromosome, 247
Yeager, Charles E., 215
Young, W. J., 17, 55
Yttrium, 3

Yukawa, Hideki, 87, 125, 142, 152, 181, 242

Z

Zero population growth (ZPG), 250
Zeus, 13
Zinc sulfide, 81, 175
Zinjanthropus, 248
Zone refining, 249
ZPG, 250
Zymase, 55